Susanne Butz

One-Dimensional Tunable Josephson Metamaterials

Experimental Condensed Matter Physics
Band 14

Eine Übersicht aller bisher in dieser Schriftenreihe erschienenen Bände finden Sie am Ende des Buchs.

One-Dimensional Tunable Josephson Metamaterials

by
Susanne Butz

KIT Scientific Publishing

Dissertation, Karlsruher Institut für Technologie (KIT)
Fakultät für Physik
Tag der mündlichen Prüfung: 11. Juli 2014
Referenten: Prof. Dr. Alexey V. Ustinov
Prof. Dr. Carsten Rockstuhl

Impressum

KIT Scientific Publishing

Karlsruher Institut für Technologie (KIT)
KIT Scientific Publishing
Straße am Forum 2
D-76131 Karlsruhe

www.ksp.kit.edu

Print on Demand 2014

ISSN 2191-9925
ISBN 978-3-7315-0271-5
DOI 10.5445/KSP/1000043318

Contents

1 Introduction **1**

2 Basic Concepts **9**

2.1 Superconductivity and Superconducting Devices 10

2.1.1 Superconductivity 10

2.1.2 The Josephson Junction 11

2.1.3 Superconducting Quantum Interference Devices 14

2.2 Transmission Line Theory 19

2.2.1 Transmission Line Basics 19

2.2.2 Scattering, Transmission and Impedance Matrix 22

2.2.3 The Periodically Loaded Transmission Line 24

2.3 Metamaterials 28

2.3.1 Maxwell's Equations and Material Parameters 28

2.3.2 Typical Meta-Atoms and their Response 30

2.3.3 Negative Index of Refraction 34

2.3.4 Material Parameters close to Zero 39

2.4 Josephson Metamaterials 41

2.4.1 The rf-SQUID as Meta-Atom 41

2.4.2 Transmission Line Setup 42

2.4.3 A Metamaterial made from Tunable, Electric Meta-Atoms 46

2.4.4 Tunable Composite Metamaterial 52

3 Experimental Setup and Procedures **55**

3.1 Sample Design and Fabrication 56

3.1.1 Fabrication 56

3.1.2 Samples 57

3.2 Experimental Setup 63

3.2.1 Mounting of the Sample 63

3.2.2 Transmission Measurement Setup 65

3.2.3 Magnetic Field Bias 67
3.2.4 Calibration Methods 69
3.3 Protecting the Samples against Stray Magnetic Flux 72
3.3.1 Abrikosov Vortices 72
3.3.2 The Effect of Stray Magnetic Fields 72

4 Experimental Results **81**
4.1 A One-Dimensional Magnetic Metamaterial 82
4.1.1 Transmission Measurements at 4.2 K 82
4.1.2 Magnetic Permeability at 4.2 K 88
4.1.3 Transmission and Magnetic Permeability at 2.1 K . . . 93
4.1.4 Discussion & Conclusion 100
4.2 A One-Dimensional Electric Metamaterial 102
4.2.1 Measurement results and Comparison to Calculations . . 102
4.2.2 CST Simulations and Comparison to Measurement . . . 105
4.2.3 Simulated Field Distribution at 12.3 GHz and 12.6 GHz . 111
4.2.4 Magnetic Flux Dependence 116
4.2.5 Discussion & Conclusion 118

5 Conclusion & Outlook **119**

Zusammenfassung & Ausblick **123**

Bibliography **127**

List of Publications **137**

Acknowledgements **139**

1 Introduction

There must be an almost infinite number of orientation week T-shirts stating

god said

$$\begin{aligned}\nabla \times \vec{E} &= -\frac{1}{\mu_0}\frac{\partial \vec{H}}{\partial t} \\ \nabla \times \vec{H} &= \varepsilon_0 \frac{\partial \vec{E}}{\partial t} + \vec{j} \\ \nabla \cdot \vec{E} &= \frac{\rho}{\varepsilon_0} \\ \nabla \cdot \vec{H} &= 0.\end{aligned} \tag{1.1}$$

and there was light.

What, to the new students, seems to be an incomprehensible inside joke at first, soon becomes the (more or less) familiar set of Maxwell's equations, that describe the propagation of electromagnetic radiation in vacuum. Finding their solution in the form of plane waves for the electric field $\vec{E}$ and the magnetic field $\vec{H}$ is a standard exercise in the electromagnetism course and a straightforward task. Maxwell's equations in material are similarly simple. This is rather surprising, since electric and magnetic fields on subatomic length scales can be very large and are very inhomogeneous both spatially and with respect to time. Any radiation passing through a medium interacts with around 10^{22} atoms per cm^3. Therefore, one would expect that the set of equations describing wave propagation through a medium becomes very large and complex. Instead it turns out that the effect of electromagnetic radiation in a material can be summarized by a frequency dependent polarization $\vec{P}$ and a magnetization $\vec{M}$ vector [Jac06]. They are obtained by applying Maxwell's equations in vacuum to individual atoms and finding the spatial average over a length scale that is short compared to the wavelength but comprises several atoms.

In an isotropic, homogeneous medium and for sufficiently weak electromagnetic fields, polarization and magnetization depend linearly on the $\vec{E}$- and $\vec{H}$-field. In this case, their effect is included in the (scalar) material parameters, namely electric permittivity $\varepsilon = \varepsilon_0\varepsilon_r$ and magnetic permeability $\mu = \mu_0\mu_r$ which replace the vacuum parameters $\varepsilon_0 = 8.85 \cdot 10^{-12}$ As/(Vm) and $\mu_0 = 4\pi \cdot 10^{-7}$ Vs/(Am) in Eq. (1.1). How ε and μ are connected to the speed of light in the medium and that the relative parameters define the index of refraction by $n^2 = \varepsilon_r\mu_r$ is well known. For dielectrics, the positive root of this equation describes the propagation of electromagnetic waves.

Although the dependence of the refractive index on material parameters has been known for a long time, only less than 50 years ago, a viable question was asked. Namely, why the negative solution for n was never considered and what new results it would yield. The man who asked that question in 1968 was Victor G. Veselago and he also gave the answer [Ves68]. By using Maxwell's equations and the boundary conditions for electric and magnetic fields at an interface, he was able to show that a negative index of refraction (NIR) is possible if the material parameters, ε and μ, are both simultaneously negative. Here, simultaneously means not only with respect to time but also with respect to frequency. Unfortunately, no such materials were and are known to exist in nature.

Materials with a negative electric permittivity on one hand are quickly found, namely in metals for frequencies below their plasma frequency. On the other hand, this task is not so easy, if not impossible for the magnetic permeability, at least in natural occurring media. It took yet another 30 years before John Pendry [Pen+99] suggested to artificially create materials, that contain engineered "atoms", designed and fabricated such that they show the requested response. This was the first idea in the field of so-called metamaterials, which should prove to be full of new and exciting physical phenomena.

Metamaterials

The "meta" in metamaterials derives from the idea that the material is composed from artificial atoms. Since these atoms, in turn, are made for normal materials - typically highly conductive metals - they are called "meta-atoms". In order to correctly use the term atom and material, the wavelength needs to be much larger than the distance between and the size of the meta-atoms, else the spatial average cannot be found. Therefore, first experiments were carried out in the microwave range ($\lambda \sim 10\,\mathrm{cm}$), using millimeter sized meta-atoms. Their fabrication is considerably simpler than constructing nanometer size atoms for the visible spectrum. Apart from the increasingly complicated fabrication, the scalability of metallic meta-atoms down to smaller size, i.e. higher frequency, is limited by Ohmic losses [Zho+05].

In 1998 and 1999, Pendry and co-workers proposed two crucial concepts. First, they suggested to use the plasmonic behavior of a metallic wire array to construct a metamaterial with negative permittivity [Pen+98].

Second, they showed how inductively coupled LC resonators can be used as magnetic meta-atoms [Pen+99]. The susceptibility of a driven oscillator around its resonance frequency behaves in a well known way. While it increases from zero to large positive values below the resonance frequency, it becomes strongly negative directly above before it approaches zero again. By combining and averaging over several of such magnetic meta-atoms, the absolute value of the averaged susceptibility may become sufficiently large to result in a negative real part of the relative magnetic permeability. The most commonly used LC oscillators for this application are so-called split ring resonators (SRR) [Pen+99; Smi+00; SSS01].

By combining SRRs with an electrically coupling wire array and operating in the band of the SRR's resonant frequency where the magnetic permeability is negative, indeed a NIR was observed experimentally [SSS01; HBC03; Par+03]. A beam of light, entering a medium with $n < 0$ from a material with positive index of refraction, e.g. vacuum, is refracted with a negative angle to the surface normal according to Snell's law for one positive and one negative index of refraction.

perfect lens

One of the most exciting applications of such a NIR metamaterial is its potential use as a "perfect lens" [Pen00]. Under the condition that the lens - a thin slab of NIR material - is impedance matched to the surrounding, positive index material, there are no reflections. Due to the negative angle of refraction, the electric field emitted from a dipole source, comes to focus twice, once inside the slab and once behind the slab. Additionally, it can be shown that evanescent waves are amplified within the NIR medium rather than decaying (energy conservation is not violated since evanescent waves do not transport energy). Hence, since both propagating and evanescent wave pass through the slab and no information is lost, resolution below the refraction limit is possible. It is however crucial, that the object is brought close to the lens, so that the evanescent component has not yet decayed too strongly.

Experimentally, subdiffraction imaging has been demonstrated in the microwave regime, using a two dimensional transmission line medium [GE04] and a patterned surface as near-field plate [GJM08] and also in the optical range, using a thin silver slab with p-polarized waves [Fan+05; Lee+05]. It should be noted that this last experiment was realized by using a material with only a negative permittivity. As was shown in Ref. [Pen00], a negative permeability is not necessary, provided that the slab is thin and p-polarized light is used.

ε- and μ-near-zero materials

Since the material parameters of any metamaterial are negative only in a specific and often narrow frequency range, there must be a frequency, where either one or both parameters pass from positive to negative (or vice versa) with a nonzero slope. When one of the material parameters is zero, so is the index of refraction at this frequency. This means, that a medium is created in which radiation has an infinite phase velocity and an infinite wave length. This effect was used to make microwaves tunnel through a short, arbitrarily shaped but narrow waveguide restriction [SE06; Edw+08]. Additionally, the phase of a wave traveling through a material with ε or μ near zero stays constant over the full length of the material [Zio04]. These properties offer a wide range of possibilities, such as phase front shaping [Zio04; Al07], phase-matching in nonlinear optics [Suc+13] and the construction of highly directive sources [Eno+02].

inhomogeneous index of refraction

Until now, we discussed metamaterials under the objective of creating media with material parameters that are (so far) unobservable in nature. Another, and not less viable approach is the realization of materials, that have a index of refraction which changes spatially in a predefined manner. One may know the experiment, in which a light beam is bent when passing through a water-sugar solution, which has a higher sugar concentration (larger n) at the bottom than at the top (lower n). The same effect is used to cloak objects. By tailoring the index of refraction in the cloaking medium surrounding the object, the path which the wave has to take is defined by Fermat's principle. In the case of proper cloaking, light exits the cloak in such a way as if the object were not there. Using this principle, a 3D cloak was used to hide a micrometer sized bump in a gold reflector in the optical range [Erg+10]. However, cloaking objects is only one small part of the larger topic of transformation optics which uses coordinate transformation in order to transfer a complex geometric problem to a simpler geometry, while mapping this coordinate transformation to new spatially inhomogeneous and anisotropic material parameters [CCS10].

tunability

Implementations of cloaking and transformation optics often employ standard dielectrics which are spatially arranged in a very specific way. This is typically possible in a comparatively large frequency range. However, the two previous implementations of metamaterials, namely NIR and n-near-zero materials are realizable only in a very narrow frequency range. For the n-near-zero materials this is because the crossover from negative to positive is at one discrete frequency. In the case of NIR materials, meta-atoms are required, that have a negative magnetic susceptibility. As mentioned above, this can realized by using inductively coupled LC resonators (SRRs). On one hand, their resonant behavior is the key ingredient, on the other it is also a very strong limitation. Since the magnetic permeability is only negative in a small frequency range directly above the resonance frequency, the NIR exists only there. In both cases, NIR and n-near-zero materials, the wanted behavior exist only in a certain frequency range and this range is fixed by the design of the meta-atoms. One way to circumvent this problem is to include an element, that renders the frequency dependent behavior of the meta-atom tunable.

There have been different approaches [Boa+11] as to how to tackle this problem. One idea, that has been investigated by different groups, involves including a varactor diode into the LC oscillator [Gil+04]. By changing the voltage applied to the varactor diode, the total capacitance of the LC oscillator, which also comprises the capacitance of the diode, is tuned and so is the resonance frequency. The voltage may be changed directly [SMK06] or by using a photodiode in the photovoltaic mode, whose voltage output is defined by the intensity of an external light source [Kap+11].

Other approaches to tunable meta-atoms include the combination of magnetic rod shaped structures [Kan+08], or the combination of liquid crystal with a magnetic metamaterial [Zha+07]. In the first case, applying an external magnetic field alters the ambient permeability of the meta-atom and hence its resonance frequency. In the second case, the resonance frequency is tuned by applying a voltage to the liquid crystals, changing the ambient permittivity.

As a last concept, the tunable kinetic inductance of a superconductor shall be mentioned. Whenever a superconductor is close to one of its critical values, which are critical temperature, critical magnetic field and critical current, its intrinsic inductance changes distinctly when either of these three parameters is changed. Ricci *et al.* demonstrated that the resonance frequency of a magnetic meta-atom made from a superconductor is tunable by changing the temperature [ROA05; RA06] and also by changing the magnetic field or driving power [Ric+07]. This concept of tunability by temperature using superconductors was also expanded from the microwave to the terahertz regime [Wu+11].

Common to all approaches that include tunable elements directly into the resonator, is the nonlinearity of this element.

Josephson Metamaterials

In this thesis, the concept and experimental realization of employing the nonlinear inductance of a Josephson tunnel contact as tunable element in a superconducting structure is presented. A Josephson tunnel contact consists of a thin insulating layer between two superconductors. Provided that the insulating layer is thin enough, a supercurrent can pass through this contact. As long as variations of the supercurrent are small, such a device behaves like a nonlinear inductor [Lic91] and its inductance depends (nonlinearly) on the current that flows through it.

By including such a Josephson junction into a superconducting loop, a radio frequency (rf-) superconducting quantum interference device (SQUID) is created. In parallel to the junction's inductance, there is also an intrinsic capacitance, which, together with the combination of loop inductance and Josephson inductance, forms an LC oscillator. Hence, like the SRR, the rf-SQUID shows resonant behavior when driven by a magnetic field. However, its resonance frequency, and thus the band in which a deviation of the magnetic susceptibility from zero is observed, is tunable due to the current dependent Josephson inductance. Using rf-SQUIDs as basic magnetic elements of a metamaterials that shows a negative permeability was theoretically suggested by Du *et al.* [DCL06; DCL08] and Lazarides *et al.* [LT07]. This thesis aims at testing the tunability of a metamaterial consisting of rf-SQUIDs and finding its potentially negative magnetic permeability. Additionally, the use of Josephson junctions in electrically interacting meta-atoms is investigated.

The tunable meta-atoms in this work are implemented in a one-dimensional (1D) coplanar waveguide geometry. Although this may defeat the notion of the system being a material at first glance, it offers several advantages. Most of the possible effects of metamaterials described above are still observable in a 1D geometry. For example, a 1D NIR metamaterial with a tunable NIR range behaves like a tunable bandpass filter, since transmission is only possible in the NIR frequency band. Another possible application may be a tunable version of a series power divider which makes use of the constant phase in an n-near-zero material [LLI05]. Additionally, the tunable 1D Josephson metamaterials in a coplanar waveguide geometry offers a potential application in the context of circuit quantum electrodynamic (cQED).

Apart from the tunability, the nonlinearity of the Josephson junction offers an additional feature, namely the occurrence of multistability [LT13] at larger driving amplitudes. This multistability including all-optical switching between different states, corresponding to different values of the susceptibility, was experimentally demonstrated by Jung *et al.* [Jun+14]. Additionally, also depending on the amplitude of the magnetic drive, the existence of localized excitations in one- and two-dimensional arrays, so-called breathers, has been predicted theoretically [ELT08].

Outlook

In the next chapter, the basics of relevant concepts for this work are explained. It starts with a brief overview of superconductivity, Josephson contacts and SQUIDs, followed by an introduction to wave propagation in coplanar waveguides. Next, the section on metamaterials gives details on the different kinds of meta-atoms and explains the occurrence of a negative index of refraction and related phenomena. The chapter ends with a detailed discussion of the electric and magnetic Josephson metamaterials investigated in this work. Chapter 3 offers a description of the different samples and their fabrication. Furthermore, it includes details on the measurement setup and procedure and discusses the measures necessary to reduce stray magnetic fields which adversely affect the performance of the metamaterial. The results of the measurements are presented, discussed and compared with calculations and simulations in Chapter 4. The last Chapter 5 finally concludes the thesis and offers an outlook.

2 Basic Concepts

In this chapter, all necessary basic theories and ideas shall be explained. It starts by covering the basic principles of superconductivity and superconducting devices such as flux quantization, the Josephson junction and superconducting quantum interference devices (SQUIDs). Next, the relevant basics of transmission line theory will be discussed. After that, we will move over to an introduction to metamaterials, which starts with a short reminder of how Maxwell's equations and material parameters govern wave propagation. It continues with a review of meta-atoms, how they can be used to achieve a negative or zero index of refraction and what interesting effects can be observed in such media. In the last part, these concepts will be put together by using superconducting devices as constituents of a metamaterial in a one dimensional transmission line setup. Thus, arriving at the central topic of this thesis, the theoretical basics and expected properties of the tunable one-dimensional Josephson metamaterial will be presented.

2.1 Superconductivity and Superconducting Devices

Superconductivity on one hand implies the transport of electrical current without dissipation [KO11], on the other hand the expulsion of any external magnetic field [MO33] and as such it obviously offers many applications. The vanishing resistance enables the construction of large coils creating magnetic fields of up to several Tesla. Its perfect diamagnetism might one day make a new superconducting design for a magnetic levitation train [Eck] possible using high temperature superconductors. In this section, however, we will concentrate on conventional low temperature superconductivity, mainly that of Niobium (Nb) and on effects arising when a weak link is formed between two superconductors.

2.1.1 Superconductivity

Superconductivity can occur when certain materials are cooled below a critical temperature. Then, electron-phonon interaction leads to the formation of so-called Cooper pairs [BCS57; Tin04] which consist of two coupled electrons with opposite spin and momentum. As a result, the Cooper pair has zero spin, making it a bosonic particle. Therefore, all Cooper pairs form a condensate, described by a single state which is energetically separated from the excited states of single electron like quasiparticles. Due to this energy gap, the Cooper pairs pass through the superconductor without scattering, thus without resistance. The wave function $\psi = \sqrt{n_s/2}\, e^{i\theta}$ describes the state of the superconducting condensate. Assuming a constant density of superconducting electrons n_s, the canonical momentum of a Cooper pair is given by

$$\vec{p} = \hbar\nabla\theta = 2m_e/(n_s e)\vec{j_s} + 2e\vec{A}. \tag{2.1}$$

Here, j_s is the supercurrent density and $\vec{A}$ the vector potential. Note that $e = 1.602{\cdot}10^{-19}$ C is the (positive) elementary charge, hence the Cooper pair carries a charge of $-2e$.

Apart from perfect conductivity, superconductivity is also accompanied by perfect diamagnetism. Any external magnetic field, that is smaller than a material and temperature dependent critical magnetic field, is expelled from the bulk material. It can penetrate the superconductor only in a small region close to the surface of the superconductor to which also superconducting currents are constrained.

flux quantization

Imagine a hole going completely through a bulk superconductor. Upon circulating once around this hole along a closed loop deep inside the superconductor, the wave function ψ, describing the superconducting condensate, has to remain single-valued. Again assuming n_s to be constant in the bulk superconductor, the phase change, when going around the closed loop once, has to be $\Delta\theta = 2\pi q$, where q is an integer. Using Eq. (2.1), this condition can be written as

$$\oint \nabla\theta d\vec{x} = \frac{1}{\hbar}\oint \vec{p}\cdot d\vec{x} = 2\pi q. \tag{2.2}$$

As mentioned before, currents in a superconductor are always located at its surface. Hence, the supercurrent density $\vec{j}_s$ vanishes, since the integration path lies deep inside the superconductor. This leads to

$$qh/(2e) = \oint \vec{A}\cdot d\vec{x} = \Phi. \tag{2.3}$$

Therefore, the magnetic flux Φ enclosed in the hole is quantized with the flux quantum $\Phi_0 = h/(2e)$ [Sch97].

type I and type II superconductivity

A superconductor for which the superconducting state vanishes completely, either when the temperature exceeds the critical temperature or when the magnetic field rises above a critical magnetic field, is called a type I superconductor. On the other hand, a type II superconductor has two critical magnetic fields. While perfect superconductivity prevails in the bulk for low magnetic fields $H < H_{c1}$, magnetic flux starts to penetrate the superconductor above a critical field H_{c1} [Abr57; Sch97] in tube-shaped non-superconducting regions. Each of these regions carries the flux of one flux quantum. Such a magnetic field penetration is called an Abrikosov vortex. The density of these vortices increases with increasing magnetic field until, at a critical value H_{c2}, superconductivity vanishes completely.

2.1.2 The Josephson Junction

When a superconductor is interrupted by a weak link, which locally suppresses superconductivity, a so-called Josephson junction is created. If the region of suppressed superconductivity is sufficiently thin, such a weak link is capable to carry a supercurrent that is sustained by tunneling Cooper pairs.

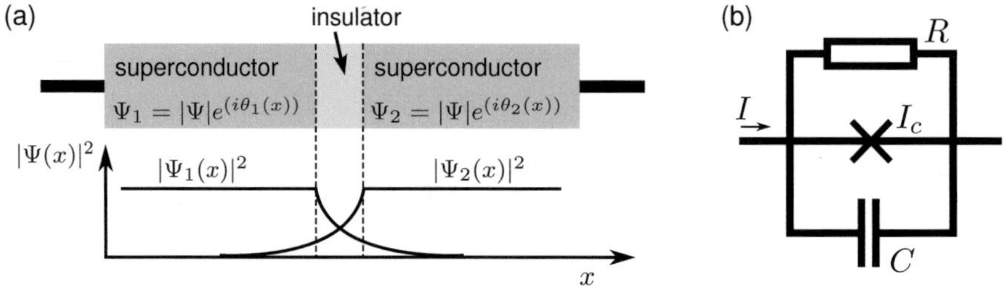

Fig. 2.1: a) Top: Sketch of a Josephson tunnel junction. Bottom: Decay and overlap of the two superconducting wave functions. b) Electric circuit equivalent of the resistively and capacitively shunted junction.

There are several possibilities of how to construct such a weak link, for example by a physical constriction of the superconducting material, or by inserting a thin non-superconducting layer between two superconducting electrodes. Here, we consider a so-called tunnel Josephson junction which uses a thin insulating layer as weak link, shown in the upper part of Fig. 2.1(a). The lower part illustrates how the superconducting wave function on either side of the barrier decays exponentially inside this weak link. The magnitude of the tunneling supercurrent is defined by the phase difference $\varphi = \theta_1 - \theta_2$ between the two wave functions on either side of the insulator. The interdependence between current through and phase difference φ across the junction is given by the first Josephson relation [Jos65]

$$I = I_c \sin\varphi. \tag{2.4}$$

Here, I_c is the critical current, i.e. the maximal value of supercurrent that can flow through the Josephson junction before it becomes resistive. It is defined by the thickness, the area and the material of the insulating layer. Once a current is applied that either exceeds the critical current or when a voltage V is applied, the junction enters the resistive state. The resistive behavior is described by the second Josephson equation [Jos65]

$$V = \frac{\Phi_0}{2\pi}\,\dot{\varphi}. \tag{2.5}$$

where the dot stands for the time derivative. In addition to the resistive channel, there is also a capacitive contribution due to the capacitance between the two superconductors. The electric circuit equivalent of this resistively and capacitively shunted junction (RCSJ) model is depicted in Fig. 2.1(b). Applying Kirchhoff's rules and Eqs. (2.4) and (2.5) to the RSCJ model, the equation of motion for the phase difference φ is

$$\frac{1}{\omega_p^2}\ddot{\varphi} + \frac{1}{\omega_c}\dot{\varphi} = j - j\sin(\varphi) \tag{2.6}$$

with the normalized bias current $j = I/Ic$, the damping defined by $\omega_c = 2\pi I_c R/\Phi_0$ and the plasma frequency $\omega_p = \sqrt{2\pi I_c/(\Phi_0 C)}$. This differential equation describes the motion of a (virtual phase) particle in a tilted washboard potential $U = -j\varphi + \cos(\varphi)$ [Sch97]. The washboard shape is due to the $\cos\varphi$, while the tilt increases with increasing bias current.

Josephson inductance

In a linear approximation, the so-called small signal approximation, the Josephson junction is interpreted as a nonlinear inductance [Lic91]: Under a small variation of the phase difference $\varphi \rightarrow \varphi_0 + \delta\varphi(t)$, Eq. (2.4) can be expanded in first order to $I(t) = I_c(\sin\varphi_0 + \delta\varphi(t)\cos\varphi_0)$. This yields an equation for $\delta\varphi(t)$ which, in turn, is inserted in Eq. (2.5).

$$\delta\varphi(t) = \frac{I(t)}{I_c\cos\varphi_0} - \frac{\sin\varphi_0}{\cos\varphi_0}, \tag{2.7}$$

$$\frac{2\pi}{\Phi_0}V(t) = \frac{d\varphi(t)}{dt} = \frac{d\varphi_0}{dt} + \frac{d\delta\varphi(t)}{dt}. \tag{2.8}$$

Taking into account the constant nature of φ_0, we arrive at

$$V = \frac{\Phi_0}{2\pi I_c\cos\varphi_0}\frac{dI}{dt}, \tag{2.9}$$

which defines a nonlinear inductor, the so-called Josephson inductance

$$L_j = \frac{\Phi_0}{2\pi I_c\cos\varphi} \tag{2.10}$$

(from here on $\varphi_0 = \varphi$). The value of this inductance is tunable by applying a dc current since the phase difference and the bias current are connected via the first Josephson equation (Eq. (2.4)). Note that this inductance becomes negative for $2\pi n + \pi/2 < \varphi < 2\pi n + 3\pi/2$ and approaches $\pm$infinity for $\varphi \approx (2n+1)\pi$, i.e. $I \approx \pm I_c$.

2.1.3 Superconducting Quantum Interference Devices

When a Josephson junction is inserted to interrupt a closed superconducting loop, a so-called rf-SQUID (radio frequency superconducting quantum interference device) is created. The prefix "radio frequency" originates from being used in the radio frequency range historically. Nowadays, possible frequencies range from MHz to several 100 GHz.

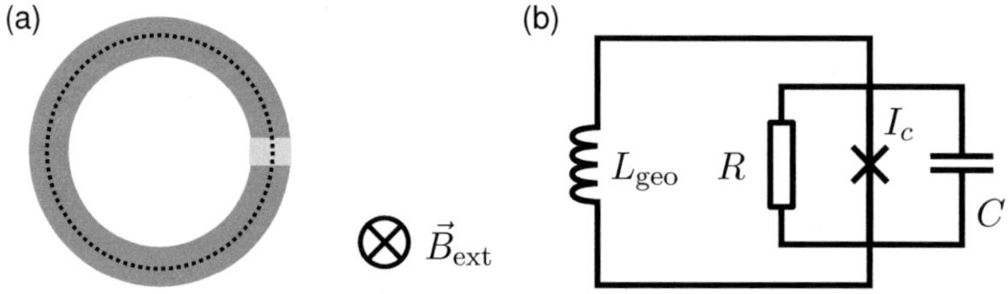

Fig. 2.2: (a) Sketch of an rf-SQUID. The darker and lighter gray areas symbolize superconducting and insulating material, respectively. The size of the areas is not drawn to scale. The dashed black line indicates the path of integration referred to in the text. (b) Equivalent circuit of the rf-SQUID using the RCSJ model.

A sketch of such a device is presented in Fig. 2.2(a). Its interesting properties become apparent once an external magnetic field $\vec{B}_{\text{ext}}$ is applied perpendicular to the area of the SQUID loop. This results in an external magnetic flux $\Phi_{\text{ext}} = \int \vec{B}_{\text{ext}}\, d\vec{S}$, where $\vec{S}$ is the surface normal of the area S of the SQUID loop.

From Sec. 2.1.1, we remember that due to the condition of a single-valued wave function, the phase change when going once around a superconducting ring has to be $\Delta\theta = 2\pi q$. This condition is weakened since the supercurrent density $\vec{j}_s$ does not vanish in the Josephson junction along the path of integration. Therefore, instead of omitting the contribution of $\vec{j}_s$ completely, an additional term in the loop integral (Eq. (2.3)) along the dashed line in Fig. 2.2(a) has to be taken into account. Furthermore, the thickness of the junction is considered to be small compared to the circumference of the loop [Jos65; SZ67].

$$2\pi q = \frac{2e}{\hbar} \oint \left(\frac{m_e}{e^2 n_s} \vec{j}_s + \vec{A} \right) d\vec{x} \tag{2.11}$$

$$\approx \frac{2e}{\hbar} \int_{\text{junction}} \left(\frac{m_e}{e^2 n_s} \vec{j}_s + \vec{A} \right) d\vec{x} + \frac{2e}{\hbar} \oint \vec{A} d\vec{x} \tag{2.12}$$

$$= \arcsin \frac{I}{I_c} + \frac{2e}{\hbar} \Phi \tag{2.13}$$

For the last step, the gauge invariant definition of the phase difference using the canonical momentum (Eq. (2.1))

$$\varphi = \frac{1}{\hbar} \int_{\text{junction}} \vec{p}\, d\vec{x} \tag{2.14}$$

and subsequently Eq. (2.4) is employed. Together with $\sin(2\pi + x) = \sin x$, we find the relation between phase and flux to be $\varphi = 2\pi\Phi/\Phi_0$. This means that the phase difference across the junction is proportional to the total flux enclosed in the rf-SQUID. However, it is important to note that the flux in the rf-SQUID is not identical to the externally applied flux Φ_{ext}. Instead, Φ_{ext} and Φ are related by the self consistent equation

$$\Phi = \Phi_{\text{ext}} - L_{\text{geo}} I(\Phi). \tag{2.15}$$

Here, L_{geo} is the geometric inductance of the SQUID loop. The induced flux, $L_{\text{geo}}I$, counteracts the external flux, thus, in this notation $I = I_c \sin\varphi$, with $I_c > 0$. From Eq. (2.15), the equation of motion of the phase particle for an rf-SQUID is obtained using the RCSJ model (Eq. (2.6)).

$$\varphi + \frac{2\pi I_c L}{\Phi_0}\left(\sin\varphi + \frac{1}{\omega_p^2}\ddot{\varphi} + \frac{1}{\omega_c}\dot{\varphi}\right) = \varphi_{\text{ext}} \tag{2.16}$$

The plasma frequency ω_p and the damping $1/\omega_c$ are defined as for Eq. (2.6). The external flux is normalized such that $\varphi_{\text{ext}} = 2\pi\Phi_{\text{ext}}/\Phi_0$. This notation for magnetic flux Φ and normalized flux φ will be valid throughout this work. The prefactor $2\pi I_c L_{\text{geo}}/\Phi_0$ gives the ratio of geometric to zero flux ($\varphi = 0$) Josephson inductance L_{j0} and will be referred to as β_L. In this work, only rf-SQUIDs with $\beta_L < 1$ will be considered. This mainly means that the dependence of Φ on Φ_{ext} (Eq. (2.15)) is unique (nonhysteretic) [Lic91].

Let us now consider what happens once the external flux contains a constant and an oscillating component $\varphi_{\text{ext}}(t) = \varphi_{\text{e0}} + \varphi_{\text{ea}}\cos(\omega t)$. For small driving amplitudes, the constant and the oscillating external flux components can be considered separately. Additionally, in the small signal limit, the behavior of the Josephson junction is that of a tunable inductor (cf. Eq. (2.9)). The constant flux component defines a constant circulating current which, in turn, determines the value of the Josephson inductance. The flux dependence of the Josephson inductance $L_{\text{j}}(\Phi_{\text{e0}})$ is obtained by solving the self-consistent equation Eq. (2.15) for Φ_{e0} and inserting the obtained solution for φ into Eq. (2.10). On the other hand, under the influence of the oscillating flux component, the rf-SQUID behaves like a driven oscillator with a resonance frequency $\omega_0 = 1/\sqrt{L_{tot}C}$. Its equivalent electric circuit for small driving amplitudes is shown in Fig. 2.2(b). The capacitance C comprises the intrinsic capacitance of the junction and any additional shunt capacitance, which decreases the resonance frequency. The total inductance is given by $L_{tot} = L_j L_{geo}/(L_j + L_{geo})$, which, due to the Josephson inductance, is tunable by the constant external flux component Φ_{e0}. This leads to a tunable resonance frequency ν_0 of the rf-SQUID:

$$\nu_0(\Phi_{e0}) = \frac{1}{2\pi\sqrt{L_{tot}(\Phi_{e0})C}} \tag{2.17}$$

Using Eq. (2.15), the dependence of the resonance frequency on external flux is calculated and plotted in Fig. 2.3 for several periods of Φ_0. Parameters typical for the rf-SQUIDs used in the experiments are used for the calculation. The periodicity in integer numbers of flux quanta is due to flux quantization as discussed earlier.

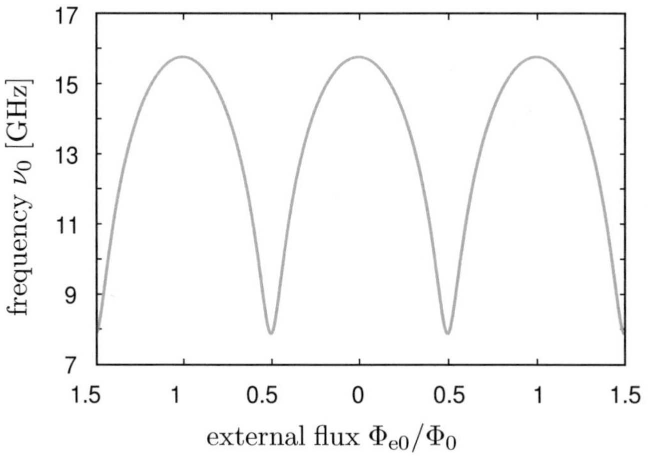

Fig. 2.3: Resonance frequency of an rf-SQUID in dependence of the external magnetic flux. The SQUID parameters are comparable to the parameters used in the experimental work: $I_c = 2.4\,\mu$A which translates to a zero flux Josephson inductance of $L_{j0} = 137$ pH. The geometric inductance is $L_{geo} = 83$ pH and the total capacitance $C = 2$ pF.

direct current (dc-) SQUID

So far, we considered a SQUID made of a superconducting loop interrupted by one Josephson junction. A so-called direct current (dc)-SQUID is formed once the loop is interrupted by two junctions. As the name already suggests, it is typically used in the direct current regime. Therefore, it is connected to a current source via two leads that are arranged such that the two Josephson junctions are in parallel. A sketch of the dc-SQUID is shown in the inset of Fig. 2.4. The boxed crosses indicate the Josephson junctions (full RCSJ model). Here, the two junctions are considered to be identical, each with a critical current I_c.

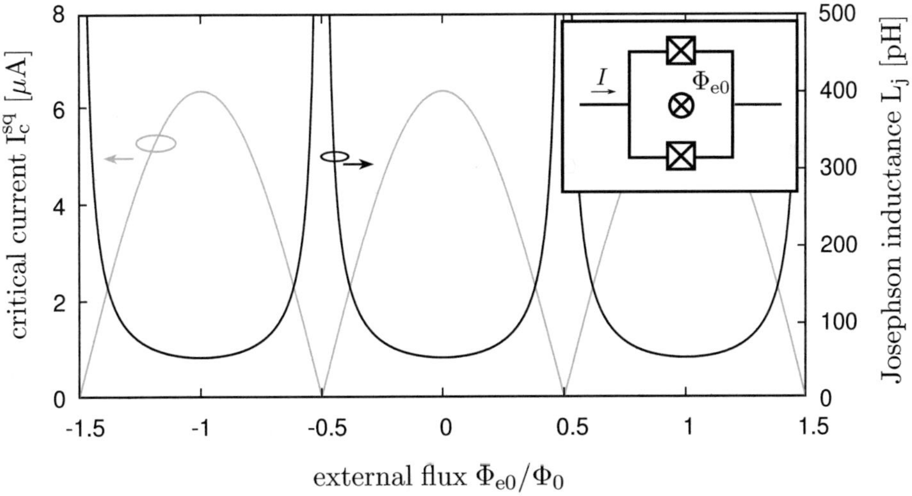

Fig. 2.4: Flux dependence of the effective critical current and Josephson inductance of a dc-SQUID with a loop inductance that is negligible compared to the Josephson inductance. The critical current of the single junction is $I_c = 3.2\,\mu$A. The inset shows a sketch of the dc-SQUID. The crosses indicate the Josephson junctions in the RCSJ model.

As long as the geometric inductance of the two arms is small compared to the Josephson inductance, it can be considered like a single junction with magnetic flux tunable critical current [CB04]

$$I_c^{sq} = 2I_c \cos(2\pi\Phi/\Phi_0). \tag{2.18}$$

In Fig. 2.4, the critical current I_c^{sq} is depicted for the case of a negligible loop inductance, i.e. $\Phi \approx \Phi_{e0}$, by the gray curve. This means that the effective

Josephson inductance of this device does not only depend on the dc current applied to it, but also on magnetic flux. This creates yet another element which acts as a flux-tunable inductance and will be used as such in the superconducting electric meta-atoms which will be introduced later in this chapter. The flux dependence of the Josephson inductance of the full dc-SQUID is shown by a black line in Fig. 2.4. For the calculation, the loop inductance is assumed to be small compared to the Josephson inductance. Note the singularity of the inductance at odd integer numbers of half flux quanta.

2.2 Transmission Line Theory

The propagation of light in free space or in media always follows a path that fulfills the principle of Fermat, which, in isotropic, homogeneous media, is a straight line. In addition, a ray of light widens and looses part of its energy during propagation due to scattering. Since, in many cases, both these properties are disadvantageous for applications, waveguides are used to control the path of propagation. Additionally, it guides the wave with as little loss as possible. Typical examples are optical fibers in the visible frequency range or so called transmission lines for microwaves. The latter will be the focus of this section although most ideas can be applied to other frequency ranges as well.

Note that for a transmission line the length in propagation direction is typically much larger than the wavelength (consider for example optical fibers used in submarine communications cable). Characteristic parameters are therefore defined per unit length.

2.2.1 Transmission Line Basics

In order to fulfill its purpose, a waveguide has to be constructed such that it restricts the fields attributed to the microwave in the space close to and around it. One way to accomplish that, is to confine the electric field between a conductor at oscillating potential V and one which is grounded. The corresponding current that oscillates in the not-grounded conductor is responsible for the magnetic field.

An example of such a waveguide including magnetic (circular) and electric (radial) field lines is shown in Fig. 2.5(a), in this case a coaxial cable. Another kind of waveguide relevant in this work is the coplanar waveguide (CPW) (cf. Fig. 2.5(b)). In principle, it is the central slice of a coaxial line (sliced along the cylindrical axis) with extended ground planes. Both behave similar, however, the field distribution in the CPW is no longer cylindrically symmetric. Nonetheless, in both waveguides the propagation mode is a so-called quasi transversal electromagnetic (QTEM) mode. In a QTEM mode, electric and magnetic field vectors both lie in the plane perpendicular to the direction of propagation. Other modes are the transversal magnetic (TM) mode, where the projection of the electric field onto the propagation direction does not vanish, and the transversal electric (TE) mode, where it is vice versa. Both modes exist in the coaxial cable and the CPW only as higher order effect above a cutoff frequency.

As will become clear later on, the QTEM mode is advantageous for the theoretical treatment and the setup of the experiments.

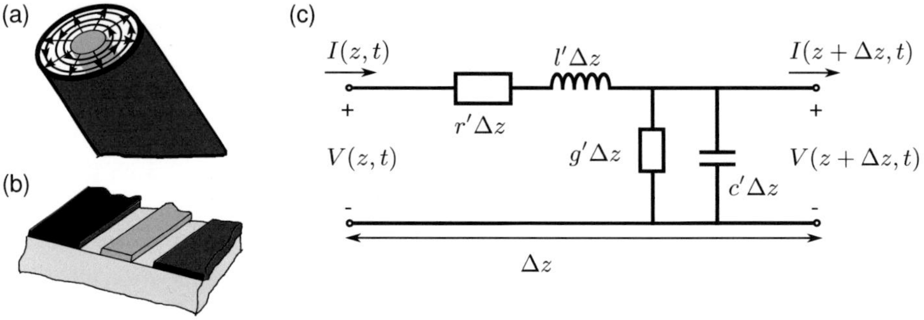

Fig. 2.5: (a) Sketch of a coaxial cable. The central conductor is shown in gray, the grounded outer jacket is black. Magnetic field lines (circular) and electric field lines (radial) are symbolically included. (b) Sketch of a coplanar waveguide. Central conductor and ground are again shown in gray and black, while the substrate is drawn in light gray. (c) Lumped-element model of a transmission line [Poz05].

The related electric quantities for magnetic and electric field are the current in the conductor and voltage between conductor and ground, respectively. Current and voltage are, in turn, related to the characteristic inductance l' and capacitance c' per unit length. Losses in the conductor and the dielectric supporting the electric field are represented by r' and g', respectively, also per unit length. A sketch of this equivalent circuit of length Δz is depicted in Fig. 2.5(c). The wave propagates in z-direction.

Using the characteristic parameters, voltage and current in the transmission line are related by the telegrapher's equations [Poz05].

$$\frac{\partial V(z,t)}{\partial z} = -r'I(z,t) - l'\frac{\partial I(z,t)}{\partial t} \qquad (2.19)$$
$$\frac{\partial I(z,t)}{\partial z} = -g'V(z,t) - c'\frac{\partial V(z,t)}{\partial t}$$

Assuming a sinusoidal signal, the solution to the telegrapher's equations at a fixed time yields current and voltage as

$$V(z) = V_0^+ e^{-\gamma z} + V_0^- e^{\gamma z}, \qquad (2.20)$$
$$I(z) = I_0^+ e^{-\gamma z} + I_0^- e^{\gamma z}, \qquad (2.21)$$

where

$$\gamma = \alpha + i\beta = \sqrt{(r' + i\omega l')(g' + i\omega c')} \tag{2.22}$$

is the complex propagation constant. V_0^+ is the voltage amplitude of the forward propagating wave, while V_0^- represents the amplitude of the reflected wave. The same is valid for the current amplitudes. It should be noted that Eqs. (2.20) and (2.21) represent the solution at a fixed point of time. Substituting Eqs. (2.20) and (2.21) back into the telegrapher's equations yields the characteristic impedance Z_0 for the forward propagating wave:

$$Z_0 = \frac{V_0^+}{I_0^+} = \sqrt{\frac{r' + i\omega l'}{g' + i\omega c'}}. \tag{2.23}$$

For the reflected wave, the impedance is defined by the respective voltage V_0^- and current I_0^-. The networks investigated in this work are all reciprocal, therefore impedances for forward and backward traveling waves are identical.

Equation (2.22) shows that the propagation constant is in general a complex number. However, often losses are small and can be neglected. This means $\alpha = 0$. Then the propagation constant and impedance are given by

$$\gamma = i\beta = i\omega\sqrt{l'c'}, \tag{2.24}$$

$$Z_0 = \sqrt{\frac{l'}{c'}}. \tag{2.25}$$

From β, the wavelength λ and the corresponding phase velocity v_p of the wave are found to be

$$\lambda = \frac{2\pi}{\beta}, \tag{2.26}$$

$$v_\mathrm{p} = \frac{\omega}{\beta} = \frac{1}{\sqrt{l'c'}}. \tag{2.27}$$

The group velocity is given as $v_\mathrm{g} = d\omega/d\beta$.

When changing back from current and voltage to magnetic and electric field, the propagation constant and the wave impedance can be found in terms of material parameters. Since the solutions for the two fields of the QTEM mode in the waveguide have the same form as a plane wave in a dielectric material, the propagation constant is given by [Poz05]

$$\beta = \omega\sqrt{\mu\varepsilon} \tag{2.28}$$

and similarly the wave impedance, which is defined by ratio of electric and magnetic field

$$Z_w = \sqrt{\frac{\mu}{\varepsilon}}. \tag{2.29}$$

Comparing Eq. (2.28) to Eq. (2.24), shows the connection between transmission line parameters l' and c' and material parameters μ and ε.

2.2.2 Scattering, Transmission and Impedance Matrix

If the impedances at a connection of two waveguides are not the same, reflections occur, comparable to the case, when light passes into a material with different index of refraction. In order to avoid this as well as possible, transmission lines are typically matched to a predefined characteristic impedance, which is either 50 Ω or 75 Ω. For semi-rigid or flexible coaxial cables used in microwave electronics, and hence in our setup, the characteristic impedance is $Z_0 = 50\,\Omega$. However, having a perfectly matched network is not always possible and scattering occurs. Thus, it is important to find a way to connect currents and voltages at each port of a multi-port network.

impedance matrix

One possibility of describing such a microwave network is by knowledge of the impedances which relate voltages and currents at the different ports by [Poz05]

$$Z_{ij} = \left.\frac{V_i}{I_j}\right|_{I_k=0 \text{ for } k\neq j}. \tag{2.30}$$

The last index means that all other ports except port j are open circuited, i.e. there is no current flowing. The inverse of the impedance matrix is called the admittance matrix $Y = Z^{-1}$.

scattering matrix

Another way of describing a microwave network is by comparing the reflected wave V_i^- at port i with the driving voltage V_j^+ of port j. In order to avoid interference due to reflections, all other ports except the driving port are terminated with a matched load. Thus, the elements of the so-called scattering matrix are

$$S_{ij} = \left. \frac{V_i^-}{V_j^+} \right|_{V_k^+=0 \text{ for } k \neq j} . \tag{2.31}$$

The scattering matrix is the quantity which is easiest accessible in measurement, for example with a vector network analyzer. Of interest in this work are only two-port networks. For this case the scattering matrix simplifies to

$$\begin{pmatrix} V_1^- \\ V_2^- \end{pmatrix} = \begin{pmatrix} S_{11} & S_{12} \\ S_{21} & S_{22} \end{pmatrix} \begin{pmatrix} V_1^+ \\ V_2^+ \end{pmatrix} . \tag{2.32}$$

By definition, S_{11} describes the reflections at port 1 and S_{21} is the transmission coefficient from port 1 to port 2, while port 2 is terminated with a matched load. Due to reciprocity $S_{21} = S_{12}$.

ABCD matrix

As a third option of describing a two-port network, the so-called transmission or ABCD matrix is often used. It relates current and voltage at the output of the network to current and voltage at the input:

$$\begin{pmatrix} V_1 \\ I_1 \end{pmatrix} = \begin{pmatrix} A & B \\ C & D \end{pmatrix} \begin{pmatrix} V_2 \\ I_2 \end{pmatrix} . \tag{2.33}$$

Therefore, the transmission line parameters, i.e. impedances and admittances need to be known. For a large network this may seem like a difficult task but any two-port network can be divided into a series of smaller and thus simpler building blocks. Now, the advantage of the ABCD matrix formalism comes to light. Since voltage and current at the $(n + 1)$-th port are calculated from voltage and current at port n, the same can be done for port $n + 2$ by means of port $n+1$. Hence, the ABCD matrix of two-port network, consisting of a series of simple building blocks, is found by simply cascading the ABCD matrices corresponding to the individual building blocks.

The ABCD matrices for common building blocks of transmission line circuits as well as different possibilities of how model a simple $50\,\Omega$ matched transmission line are given in literature [Poz05]. Note, that in order to avoid confusion with the transmission coefficient S_{21} of the scattering matrix which describes the transmission from port 1 to port 2, always the term ABCD matrix will be used instead of transmission matrix.

Since impedance, scattering and ABCD matrices all describe the same system, they can be converted into each other. The conversion formulas between different matrices can be found in literature [Poz05] and are not given here.

2.2.3 The Periodically Loaded Transmission Line

Consider an infinitely long, reciprocal transmission line that is periodically loaded with admittances Y_3 as sketched in Fig. 2.6. The unloaded line is described by the impedance $Z_0 = \sqrt{l'/c'}$ and the propagation constant $k = \omega/c$, where $c = 1/\sqrt{l'c'}$ is the speed of light in the unloaded line.

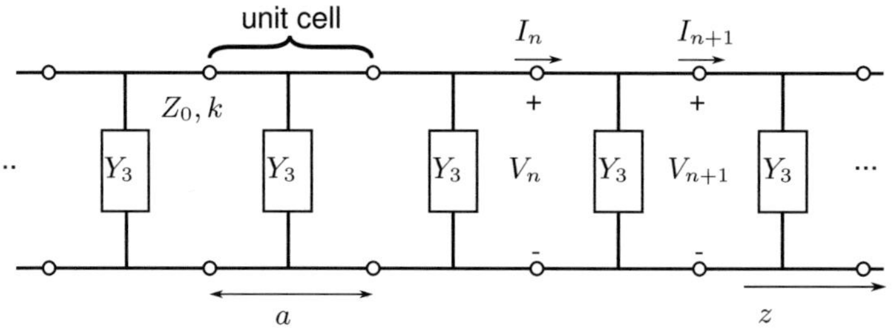

Fig. 2.6: Sketch of a periodically loaded transmission line [Poz05].

A wave propagating in $+z$-direction is given by

$$V(z) = V(0)\, e^{-\gamma z}, \tag{2.34}$$

$$I(z) = I(0)\, e^{-\gamma z}. \tag{2.35}$$

The ABCD matrix of one unit cell of length a

$$\begin{pmatrix} A & B \\ C & D \end{pmatrix} = \tag{2.36}$$

$$\begin{pmatrix} (\cos ka + \frac{b}{2}\sin ka) & iZ_0(\sin ka - \frac{b}{2}\cos ka + \frac{b}{2}) \\ iZ_0(\sin ka - \frac{b}{2}\cos ka - \frac{b}{2}) & (\cos ka + \frac{b}{2}\sin ka) \end{pmatrix},$$

with $b = iY_3Z_0$, is used to connect voltage and current before and after one unit cell. Note, that this definition of b differs from the definition in Ref. [Poz05]. As a result, a relation between the propagation constant of the unloaded line k, and the propagation constant of the wave in the loaded line $\gamma = \alpha + i\beta$ and thus the dispersion relation is found [Poz05]:

$$\cosh \gamma d = \cos kd + \frac{b}{2} \sin kd. \tag{2.37}$$

In the case, that the admittance Y_3 is purely imaginary, b can be written as $b = -\mathrm{Im}(\mathrm{Y3})\mathrm{Z_0}$ which is purely real. Then, the left hand side of this equation also has to be purely real. Therefore, γ has to be either purely imaginary ($\gamma = i\beta$) or purely real ($\gamma = \alpha$). According to the definition of the cosh

$$|\cosh \gamma a| = \begin{cases} |\cos \beta a| & \leq 1, \qquad \gamma = i\beta \\ |\cosh \alpha a| & \geq 1, \qquad \gamma = \alpha. \end{cases}, \tag{2.38}$$

frequency bands with propagating waves ($\gamma = i\beta$), so-called pass bands, are only possible at frequencies for which the absolute value of the right hand side of Eq. (2.37) is less than unity. Thus, they are found by solving

$$|\cos \frac{\omega}{c} d + \frac{b}{2} \sin \frac{\omega}{c} d| - 1 = 0 \tag{2.39}$$

with $b = -\mathrm{Im}(\mathrm{Y3})\mathrm{Z_0}$. From Eq. (2.38), the upper cutoff frequencies of the pass bands are extracted by finding the frequencies for which β is equal to integer numbers of π/a. The opening of the stop bands at these β-values is a well known property of Bloch waves [Kit05], i.e. any kind of waves propagating through any kind of periodic structure, as nicely pointed out in Refs. [Bri53] and [RBLL95]. For example, Eq. (2.37) is also the result of the Kronig-Penning model, which considers electron propagation through a periodic delta-function potential [Kit05].

The reason for so-called stop bands, i.e. frequency bands where no wave propagation is possible, is easily found. At propagation constants $\beta = q\pi/a$, $q \in \mathbb{Z}$, integer numbers of half wavelengths are equal to the unit cell length a. Incoming and reflected wave fulfill the Bragg condition, which creates a standing wave. Depending on whether the superposition of incoming and reflected wave is symmetric or antisymmetric, the resulting standing wave either decreases or increases the energy of the system. This leads to an energy (frequency) gap between $\beta = (\pi/a)^-$ and $\beta = (\pi/a)^+$. Here $^{+/-}$ stands for the limes from below and above, respectively.

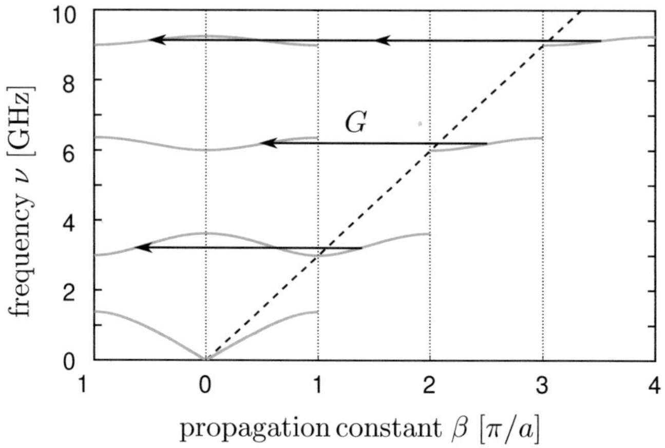

Fig. 2.7: Dispersion relation of a infinite transmission line loaded periodically with capacitances, such that $Y_3 = i\omega C$ with $C = 2.6\,\mathrm{pF}$, $a = 5\,\mathrm{cm}$. The dashed line is given by the dispersion line of the unloaded line ($Y_3 = \infty$) $\omega = \beta c$.

Staying within the concept of Bloch waves, the full zone scheme of the dispersion relation $\omega(\beta)$ for an infinite transmission line that is periodically loaded with a capacitance C is shown in Fig. 2.7, in addition to its reduction to the first Brillouin zone. The length of the unit cell is $a = 5\,\mathrm{cm}$. The black arrows indicate how the reduced zone scheme is constructed by subtracting integer numbers of reciprocal lattice vectors $G = 2\pi/a$. Considering only the reduced zone scheme ($-1 < \beta a/\pi < 1$), phase and group velocity have the same sign in the first and third pass band. The signs of the velocities are opposite in the second and forth pass band. This is due to the discreteness of the lattice which leads to an ambiguity in possible wavelengths [Bri53]. Namely, the pass band between $1 < \beta a/\pi < 2$ can be expressed either as a wave with $\beta a/\pi > 1$ and thus $\lambda < 2a$ or with $-1 < \beta a/\pi < 0$ and $\lambda > 2a$.

Bloch impedance

Since the periodic load of a transmission line strongly affects the wave propagation, this has to be reflected also in its impedance. The so-called Bloch impedance Z_B is the characteristic impedance at the terminals of the unit cell. Using the ABCD matrix of one unit cell Eq. (2.36), it is calculated to be [Poz05]

$$Z_\mathrm{B}^{\pm} = \frac{\pm B Z_0}{\sqrt{A^2 - 1}}, \tag{2.40}$$

where A and B are defined by Eq. (2.36). Again, $\pm$ indicates positively or negatively traveling waves. If B is purely imaginary, for exmpale in the case for $b = -\mathrm{Im(Y3)Z_0}$, Z_B is real for $A \leq 1$. This corresponds, as expected, to the $\cos$ solution of the $\cosh$ (cf. Eq. (2.38)).

finite length aspects

Due to the infinite number of unit cells in the transmission line, the dispersion relation in the pass band is continuous as seen in Fig. 2.7. However, if the number of unit cells is finite, transmission is possible only for discrete frequencies. Simplified, due to the impedance mismatch between $50\,\Omega$ and Z_B of the periodically loaded transmission line of finite length, its behavior is comparable to a transmission line resonator. Wave propagation is only possible when integer numbers of $\lambda/2$ fit into the loaded transmission line. It should be noted that due to the nonlinear dependence $\omega(\lambda)$ the allowed frequencies are not equally spaced, different from a transmission line resonator.

2.3 Metamaterials

A metamaterial is a medium constructed from artificial elements, so-called meta-atoms, with the purpose to manipulate wave propagation through it in a desired fashion. As in natural occurring media, the distance between and the size of these meta-atoms in propagation direction are small compared to the vacuum wavelength of the incoming wave [SPW04]. Typically, this is still valid for the propagation in the medium unless the absolute value of the index of refraction is much larger than unity. Then, the wavelength may be shortened to be on the order of the meta-atom size and periodicity. Such a system usually falls under the term photonic crystals, i.e. media which are also made of artificially engineered structures however with a periodicity on the order of the wavelength.

Provided that the above size restrictions are fulfilled and the term metamaterial is applicable, such an artificial material can be treated as an effective medium. Instead of evaluating the interaction of radiation with each meta-atom separately, the average response of several meta-atoms is considered. Consequently, the propagation of electromagnetic radiation is then described using the material parameters electric permittivity and magnetic permeability.

2.3.1 Maxwell's Equations and Material Parameters

Maxwell's equations in vacuum were already presented in the introduction (cf. Eq. (1.1)). In the following, Maxwell's equation in media will be discussed in more detail. In an isotropic, homogeneous and linear medium, polarization and magnetization are given by $\vec{P} = \varepsilon_0 \chi_e(\omega)\vec{E}$ and $\vec{M} = \chi_m(\omega)\vec{H}$. The scalar susceptibilities are used to define the material parameters magnetic permeability μ and electric permittivity ε (the latter is also known as dielectric constant).

$$\mu(\omega) = \mu_0(1 + \chi_m(\omega)) = \mu_0\mu_r(\omega) \tag{2.41}$$

$$\varepsilon(\omega) = \varepsilon_0(1 + \chi_e(\omega)) = \varepsilon_0\varepsilon_r(\omega) \tag{2.42}$$

The frequency dependence of the material parameters will be of importance later and should be kept in mind, although it will omitted in the following equations for clarity.

Using the above considerations, Maxwell's equation in media for a fixed frequency are given by

$$\nabla \times \vec{E} = -\frac{\partial}{\partial t}\vec{B} \quad (2.43)$$

$$\nabla \times \vec{H} = \frac{\partial}{\partial t}\vec{D} + \vec{j} \quad (2.44)$$

$$\nabla \cdot \vec{D} = \rho \quad (2.45)$$

$$\nabla \cdot \vec{B} = 0 \quad (2.46)$$

for electric field $\vec{E}$ and magnetic field $\vec{H}$ and taking into account charge and current density ρ and $\vec{j}$, respectively. The dielectric displacement $\vec{D} = \varepsilon\vec{E}$ and the magnetic induction $\vec{B} = \mu\vec{H}$ will be of use later on.

In a dielectric material at optical frequencies, the relative permittivity ε_r is typically larger than unity, while $\mu_r = 1$. It should be noted that, in general, the relative material parameters are complex numbers, but unless stated otherwise they are considered to be real in this section for simplicity. In this case, the index of refraction, which is defined by

$$n^2 = \varepsilon_r\mu_r, \quad (2.47)$$

is a real number. The speed of light in the medium is given by

$$c = \frac{1}{\sqrt{\varepsilon\mu}}. \quad (2.48)$$

Note the similarity to the definition of c in a transmission line as given in Sec. 2.2.1. A plane wave propagating with speed c through a medium with an refractive index n is described by $e^{i\omega t - i\vec{k}\vec{r}}$. Here, $\vec{k} = n\vec{k}_0$ is the wave vector in the medium and $|\vec{k}_0| = \omega/c_0$ is the length of the wave vector in vacuum.

On the other hand, the opacity and reflectivity of metals in and below the optical spectrum is due to a negative electric permittivity (and simultaneously positive permeability). This yields an imaginary value for n, which in turns leads to an exponentially decaying wave. Therefore, radiation in and below the optical range cannot penetrate a metal. The same would be valid for a hypothetical material with $\mu_r < 0$ and $\varepsilon_r > 0$, but in nature there are no materials with $\mu_r < 0$.

Being able to influence and tune these two parameters at will is the main objective when constructing and using metamaterials. As already pointed out in the introduction, two different, although connected, motivations are distinguishable. On one hand, metamaterials are constructed such that, due to a spatially varying index of refraction, waves are conducted along a predefined path. This concept is for example used for cloaking. On the other hand, new meta-atoms are invented that enable the creation of materials that have material parameters which are not positive and larger or equal to unity. This work is involved solely with the latter motivation. Therefore, meta-atoms have to be designed that have a susceptibility which is less than zero, i.e. that are able to counteract an incoming wave.

2.3.2 Typical Meta-Atoms and their Response

The electric and magnetic interactions of media with electromagnetic radiation are fundamentally different. There is a monopole charge corresponding to the electric field but no corresponding monopole for the magnetic field. Therefore, the coupling mechanisms and hence the approaches as to how to design meta-atoms for interaction with either electric or magnetic field component differ. As a result, the task of creating a metamaterial is strongly simplified by designing separate meta-atoms for the interaction with the electric and with the magnetic field component and combining them afterwards.

electrically interacting elements

The propagation of electromagnetic radiation through a transparent dielectric medium is defined only by the interaction with the electric field. The electronic cloud is localized and oscillates, driven by the electric field, around the atom's core, creating on oscillating dipole. The spatially averaged result is an effective wave, that travels at a speed $c = c_0/n$. The refractive index n depends only on the relative electric permeability ε_r (since $\mu_r = 1$). The relative electric permittivity, in turn, is defined by the electric susceptibility of the electronic cloud to the incoming wave.

Metals, on the other hand, have electrons that are delocalized and therefore their response to an incoming electromagnetic wave is different from that of a dielectric. Two cases have to be distinguished. Radiation with a frequency below the plasma frequency ω_p of the free electrons in the metal will be reflected, due to the moving electrons. Above the plasma frequency, however, the electrons cannot follow and screen the electric wave anymore, and the medium becomes transparent. The relative electric permittivity of metals in dependence on frequency and without considering dissipative effects is given by

$$\varepsilon_r = 1 - \frac{\omega_p^2}{\omega^2}, \tag{2.49}$$

$$\text{with} \quad \omega_p^2 = \frac{n_e e^2}{\varepsilon_0 m_e}, \tag{2.50}$$

where n_e is the electron density and m_e the electron mass. For conventional metals, the plasma frequency is in the ultraviolet region. They are therefore reflective and have a negative electric permittivity in the optical and microwave frequency range.

It seems obvious to make use of this plasmonic behavior of metals below the plasma frequency. The rather ingenious idea by Pendry et al. [Pen+98] was to remove almost all of a bulk metal and only leave a periodic array of thin wires standing. The plasma frequency of such a medium is lower than the plasma frequency of the bulk material since the effective mass of the electrons increases for two reasons. First, due to the wire structure the inductance is increased and second, the average effective electron density decreases due to the loss of bulk material. Hence the medium becomes transparent with $\varepsilon_r > 0$ already at lower frequencies.

The predicted behavior was successfully tested in the aforementioned initial work by Pendry as well as in various other works [Smi+00; GBMM02], for example also for superconducting wires [ROA05]. The biggest advantage of stripping most of the metallic material away is that there is space for a sub-lattice consisting of magnetically coupling meta-atoms, as also shown in Ref. [Smi+00].

magnetically interacting elements

The relative magnetic permeability of any natural dielectric material does not deviate much from unity. In fact, the only dielectric materials with $\mu_r > 1$ are artificial materials, where nanometer-size magnetic particles are embedded into a dielectric matrix. For such materials, values of μ_r up to 30 are possible [Pie+07]. Much larger permeabilities are found in magnetic materials (metals). Additionally, there are no natural materials that have a negative magnetic susceptibility (in zero external magnetic field) at high frequencies.

The approach of how to construct a meta-atom that couples to the magnetic component of an electromagnetic wave is quite different from the previously introduced wire array for the $\vec{E}$-field. Since there is no magnetic charge, there is no plasma frequency. The idea to use a resonant RLC circuit and employ its non-zero susceptibility around the resonance frequency was again first introduced by Pendry et al. [Pen+99]. A loop or cylinder with its rotational axis oriented parallel to the magnetic field $\vec{B}$ is employed together with a capacitance. The initial proposal included swiss roll cylinders and open cylinders or open rings within each other. The example of a double split ring resonator (SRR) is sketched in Fig. 2.8(a). It consists of two concentric conducting loops with different radii that are both interrupted at one point. Its capacitance is defined by the width of the gaps between the two rings and the width of the interruption.

In order to understand the response of an SRR to a sinusoidal magnetic driving field perpendicular to the loop area, it suffices to consider a simple RLC circuit as depicted in Fig. 2.8(b). Taking into account the loop area, the magnetic field is replaced by the corresponding magnetic flux threading the loop as was done in Sec 2.1.3 for the rf-SQUID. This spares us the consideration of the spatial dependence of the magnetic field.

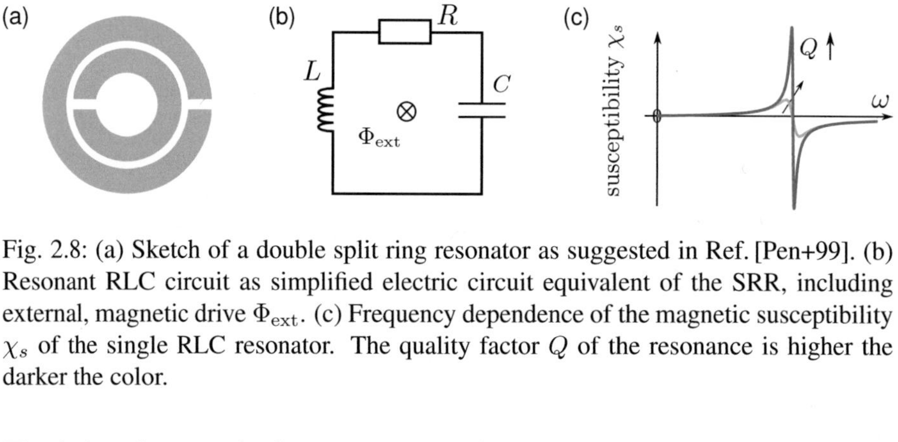

Fig. 2.8: (a) Sketch of a double split ring resonator as suggested in Ref. [Pen+99]. (b) Resonant RLC circuit as simplified electric circuit equivalent of the SRR, including external, magnetic drive Φ_{ext}. (c) Frequency dependence of the magnetic susceptibility χ_s of the single RLC resonator. The quality factor Q of the resonance is higher the darker the color.

The induced magnetic flux response of the RLC resonator to the flux drive $\Phi_{\text{ext}} = \Phi_{\text{ea}} e^{i\omega t} + c.c.$ is given by $\Phi_{\text{res}} = \Phi_{\text{ra}} e^{i(\omega t + \delta)} + c.c.$. Here, ω is the frequency of the driving signal, δ is the relative phase between incoming and induced signal and $c.c.$ stands for the complex conjugate. In a frequency range around the resonance frequency $\omega_0 = 1/\sqrt{LC}$, the amplitude of the response Φ_{ra} of the resonator is larger than the amplitude of driving field Φ_{ea}. Additionally, the relative phase δ between the two signals shifts quickly from zero around the resonance frequency and approaches π asymptotically for $\omega \to \infty$.

Combining amplitude and phase of the incoming and induced signal, the susceptibility of the single element is defined as

$$\chi_{\mathrm{s}} = \langle \frac{\Phi_{\mathrm{res}} - \Phi_{\mathrm{ext}}}{\Phi_{\mathrm{ext}}} \rangle_t . \tag{2.51}$$

Here, $\langle\rangle_t$ stands for the time average. Its frequency dependence is depicted in Fig. 2.8(c). Around the resonance frequency, the susceptibility clearly deviates from zero and even becomes negative. Thus, a material that consists of many of these oscillators exhibits a magnetic permeability $\mu_r = 1 + \tilde{F}\chi_s$ that deviates from unity around the resonance frequency and may even become negative. $\tilde{F}$ accounts for a filling factor and depends on the geometry of the meta-atom and the spatial composition of the corresponding metamaterial. It should be noted that this definition of the relative magnetic permeability including Eq. (2.51) is valid only up to a frequency $\omega < \sqrt{3}\omega_0$ [Cap09], since μ_r does not approach unity for $\omega \to \infty$. In order to avoid this unphysical behavior, another expression for the magnetic permeability can be found [SK00], with the draw back that here unity is not reached in the low frequency limit. Since both expressions describe the behavior at resonance correctly, and the high frequency limit is not of interest in this work, we will use the given expression together with Eq. (2.51).

Apart from a large filling factor $\tilde{F}$, a high quality factor Q of the resonance is crucial to obtain a negative permeability. With increasing losses, the maximal relative amplitude and the steepness of the phase shift decrease. This, in turn, decreases the deviation of the susceptibility from zero (cf. Fig. 2.8(c)). Consequently, if dissipation in the resonator is too large, the variation of the magnetic permeability may not go below zero anymore.

There are many different designs for magnetic meta-atoms [Pen+99; Sha07], however, common to all is the resonant nature. Being unavoidable, the resonant behavior is also a strong limitation. From Fig. 2.8(c), it is obvious that a specific value of χ_s and thus a specific value for μ_r is achievable only at a fixed frequency. Additionally, the frequency range, where a negative μ_r is feasible (i.e. where $\tilde{F}\chi_s < -1$) is narrow and close to resonance, where losses in the resonator are highest.

As a side note, referring to a statement from the beginning "there are no natural materials that have a negative, high frequency magnetic susceptibility (in zero external magnetic field)": This is no longer true, once a magnetic field is applied to materials with permanent magnetic dipole moments. This magnetic field may also be caused locally within the medium by the material itself. Due to the Zeeman effect [Zee97], its degenerate energy levels, defined by the magnetic quantum number, split up. Such a material exhibits paramagnetic resonances (EPR) whenever a magnetic driving field is in resonance with one of the transition between these levels [AB70]. As for the RLC circuits, the relative magnetic permeability of such a material at resonance deviates from unity. To the author's knowledge, there has been no investigation considering the EPR in the context of metamaterials, although the effect of EPR is well understood and widely used.

2.3.3 Negative Index of Refraction

Victor G. Veselago was among the first who considered the possibility that Eq. (2.47), which defines the index of refraction, allows in fact two solutions for n, namely $n = \pm\sqrt{\varepsilon_r \mu_r}$ [Ves68]. He claimed that simultaneously negative relative material parameters ε_r, μ_r compel the selection of the negative root for the index of refraction while the positive index of refraction is valid for positive relative material parameters.

In order to verify this claim, a closer look has to be taken at Maxwell's equations in media (cf. Eqs. (2.43)-(2.46)) and the direction of electric field $\vec{E}$ and magnetic induction $\vec{B} = \mu \vec{H}$. Assuming a plane wave for the electric field $\vec{E} = \vec{E}_0 \exp(i(\omega t - \vec{k}\vec{x}))$ (and the same for the magnetic induction $\vec{B}$), together with a current and source free environment, they can be simplified to

$$\begin{aligned} \vec{k} \times \vec{B} &= -\omega \varepsilon \mu \vec{E} \\ \vec{k} \times \vec{E} &= \omega \vec{B}. \end{aligned} \tag{2.52}$$

In Fig. 2.9, an interface between a conventional material at the top with $\varepsilon_1, \mu_1 > 0$ and a material with both $\varepsilon_2, \mu_2 < 0$ at the bottom is displayed. The arrows denote the direction of $\vec{E}$- and $\vec{B}$-field and corresponding dielectric displacement $\vec{D}$ and magnetic field $\vec{H}$.

The propagation of light in an isotropic medium with positive n is well known: $\vec{E}$, $\vec{B}$ and the propagation $\vec{k}$ form a right-handed system, so does by definition the combination of $\vec{E}$, $\vec{B}$ and Poynting vector $\vec{S}$.

In order to find the behavior in a medium with simultaneously negative permittivity and permeability, the behavior of the four fields when crossing the

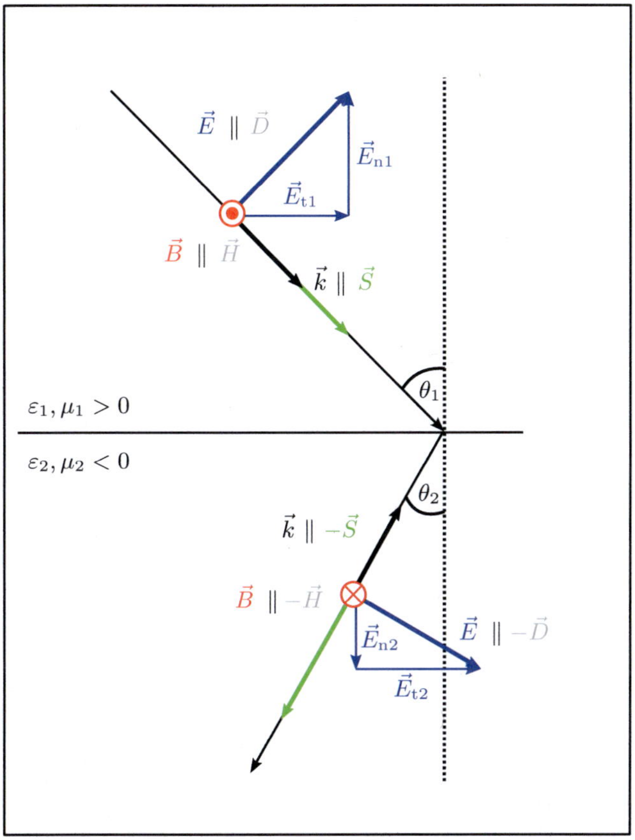

Fig. 2.9: Interface between a medium with purely positive material parameters ε_1, μ_1 (top) and one with negative material parameters ε_2, μ_2 (bottom). Both materials are isotropic. The field vectors of an electromagnetic wave under oblique incidence from the top are given as well as wave vector $\vec{k}$ and Poynting vector $\vec{S}$. When passing through the interface, the ratio of the normal components is given by the ratio of ε_1 and ε_2. Hence, $\vec{E}_{\mathrm{n1}}$ is anti-parallel to $\vec{E}_{\mathrm{n2}}$. The tangential component is not affected. Combining both components in the NIR medium yields the new $\vec{E}$-field vector. Similar considerations apply for $\vec{B}$. Note that in the NIR material $\vec{k}$ and $\vec{S}$ point in opposite direction, as do $\vec{E}$ and $\vec{D}$ as well as $\vec{B}$ and $\vec{H}$.

interface needs to be investigated. Therefore, the field vectors are split into components tangential and normal to the interface.

Due to Faraday's law of induction the parallel field component of $\vec{E}$ is continuous when crossing the interface [Jac06]. The ratio of the normal components on either side of the interface is determined by the ratio of the two different values for the permittivity $\varepsilon_1, \varepsilon_2$. Thus, after crossing from a medium with positive ε into a medium with negative ε, the normal component of the electric field points into the opposite direction. The reconstruction of the magnetic induction vector $\vec{B}$ works in a similar way.

By applying Eqs. (2.43)-(2.46) to the these newly constructed vectors, it turns out that a beam of light is refracted with an negative angle θ_2. The propagation direction $\vec{k}$ is now antiparallel to the direction of energy flow, i.e. $\vec{S}$. This also means, that $\vec{B}$, $\vec{E}$ and $\vec{k}$ form a left-handed system and that phase and group velocities have opposite signs. The group velocity v_{g}, defining the direction of energy flow, follows $\vec{S}$ while the phase velocity v_{ph} is in direction of $\vec{k}$.

Applying Snell's law to the left-handed system shows that the beam propagation in the second medium is described by a negative index of refraction. In conclusion, Veselago's claim, that negative material parameters require the choice of the negative root of Eq. (2.47), is indeed true: In order to observe a negative index of refraction, a material has to be found with simultaneously negative electric permittivity and magnetic permeability. Simultaneous in this context means not only with respect to time but also in an overlapping frequency range.

experimental realization of materials with $n < 0$

As briefly mentioned before, wave propagation through a medium with only one negative material parameter is not possible. However, at frequencies where both material parameters are negative, transmission through this negative index of refraction (NIR) material is finite as was shown in Ref. [Smi+00]. One year later, negative refraction was for the first time experimentally tested in a wedge-shaped piece of material for a frequency of 10.5 GHz [SSS01].

Since the maximal possible transmission is limited by losses in the material, they have to be reduced as much as possible. This becomes increasingly difficult as the size of the meta-atoms is decreased in order to reach higher frequencies.

Nonetheless, since the first realization in 2000, materials with negative index of refraction have been built for frequencies up to the optical range [Sha07].

For this purpose, the meta-atoms have to been miniaturized. Unfortunately, the resonance frequency of the SRR does not increase linearly with decreasing size anymore when approaching the optical frequency range. Instead it saturates, while the amplitude of the resonance decreases until $\mu_r < 0$ is no longer reachable [Zho+05]. Therefore, the shape of the meta-atoms has to be optimized [Lin+04] for the THz-frequency region and, in order to obtain a negative index of refraction in the visible range, magnetic and electric components were combined to a so-called fishnet structure [Dol+06; Dol+07].

negative index of refraction in transmission lines

Another approach, as to how a material with a negative index of refraction can be realized, was introduced in 2002 by Eleftheriades et al. [EIK02]. Instead of considering light propagation in bulk-like materials constructed from an arrangement of individual meta-atoms, they investigated the possibility of achieving a NIR in a distributed transmission line network.

Already 70 years ago, Maxwell's equation in 1D, 2D or 3D media were modeled by 1D, 2D or 3D transmission line networks [Kro44]. By comparing the equations describing electric and magnetic field (from Maxwell's equations Eqs. (2.43)-(2.46)) to the equations describing voltage and current respectively (from the telegrapher's equations Eq. (2.19)), a relation between transmission line parameters, i.e. impedance $Z = i\omega L$ and admittance $Y = i\omega C$ as illustrated for 1D in Fig. 2.10(a), and material parameters per unit cell with length a is found [EIK02].

$$\mu(\omega) = \frac{Z(\omega)/a}{i\omega} \tag{2.53}$$

$$\varepsilon(\omega) = \frac{Y(\omega)/a}{i\omega} \tag{2.54}$$

These two equations yield $Z = i\omega\mu a$ and $Y = i\omega\varepsilon a$ for one unit cell which, in turn, defines $L = \mu a$ and $C = \varepsilon a$. The proportionality of L and μ as well as C and ε leads to the conclusion, that by implementing elements with negative values for L and C, a medium with both $\mu, \varepsilon < 0$ is created.

The question is now, how negative values for inductance and capacitance can be engineered. In the telegrapher's equations, the minus sign in front of L and C leads to an effective phase shift of π between current and voltage. The same phase change would be observed, if inductance and capacitance in the transmission line were exchanged. Therefore and by also taking into account the frequency dependence of Z and Y, we can rewrite $-L = 1/(\omega^2 C_{\mathrm{lh}})$ and vice versa. A sketch of such a left-handed transmission line in 1D is given in Fig. 2.10(b). Due to the frequency dependence, the left-handed inductance is not identical to the inductance in the right handed medium and $C_{\mathrm{lh}} \neq C$. Nonetheless, by designing a transmission line from lumped elements with capacitances in series and shunt inductances, a left-handed medium can be built. It should be noted, that such a transmission line also always comprises a (small) right-handed contribution (L_{rh} and C_{rh}) due to the host transmission line. But this is of course also true for the bulk metamaterials introduced in the previous subsection.

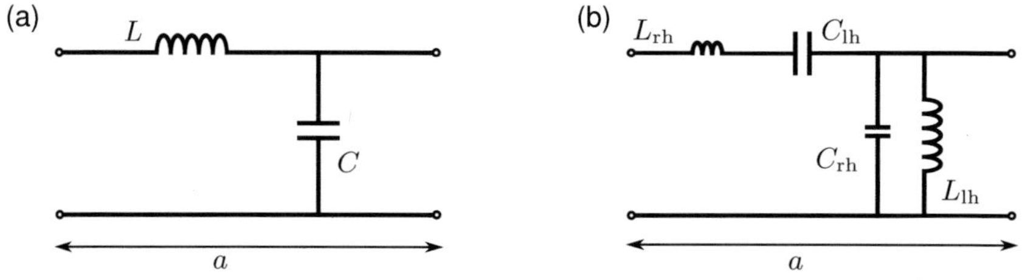

Fig. 2.10: (a) Sketch of one unit cell of a right-handed (standard) transmission line in 1D loaded with an inductance L and capacitance C. (b) Sketch of one unit cell of a left-handed transmission line. The (right-handed) host transmission line is defined by L_{rh} and C_{rh}. The (dominant) left-handed behavior is given by L_{lh} and C_{lh}.

While the NIR in 2D distributed networks has been used to demonstrate imaging with a resolution better than the diffraction limit [GE04], reducing the transmission line metamaterial to one dimension sacrifices the refractive, i.e. angle dependent properties of a NIR material. Nonetheless, the effective parameters still define the impedance and the propagation vector in the medium. One dimensional NIR transmission lines were for example used to realize compact phase-shifters [AE03] or N-port series power dividers [LLI05].

backward traveling waves vs. negative index of refraction

As previously discussed, a negative index of refraction leads to opposite signs for phase and group velocity. The same was found in Sec. 2.2.3 for the second and the forth pass band in the reduced zone scheme of a transmission line loaded periodically with capacitances C. Hence, the question may come up, what the difference between the two effects is. The answer was also already given in Sec. 2.2.3. A negative n is a (meta)material effect, which means, that the wavelength is much larger than the unit cell length. On the other hand, the opposite sign of phase and group velocity in a periodically loaded transmission line arises from an ambiguity in wave vector definition by adding or subtracting integer numbers of reciprocal lattice vectors. In terms of wavelengths, this relates to a subtraction or addition of integer numbers of unit cell length. The second pass band in the reduced zone scheme (cf. Fig. 2.7) corresponds in fact to a wave with positive phase and group velocity and with a wavelength between $a < \lambda < 2a$. Hence, the condition of $\lambda \gg a$ is no longer satisfied.

2.3.4 Material Parameters close to Zero

In this section on metamaterials, we have learned so far that the material parameters of a linear medium constructed of meta-atoms can vary between positive and negative values depending on frequency. Consequently, there is a frequency where either one or both parameters pass through zero. Using Eq. (2.47), the index of refraction at that frequency is also zero.

First, consider the case when $\mu = \varepsilon = 0$ and the electric (and magnetic) field is of the form $\vec{E}(\vec{r})\exp(i\omega t) + c.c.$ (and respective for $\vec{H}$). Under these conditions Maxwell's equations in media (Eqs. (2.43)-(2.46)) reduce to the expressions

$$\nabla \times \vec{E}(\vec{r}) = 0 \tag{2.55}$$

$$\nabla \times \vec{H}(\vec{r}) = \vec{j} \tag{2.56}$$

which describe spatially static electric and magnetic fields. This translates into an infinite wavelength, as can also be seen from $\lambda = \lambda_0/n$. On the other hand, the phase velocity is infinite. This results in rather interesting properties.

For example, it was shown using finite difference time domain (FDTD) simulations [Zio04] that the phase of a wave that travels through a slab of $n = 0$ material stays constant across the slab. This reflects both infinite wavelength and infinite phase velocity. Although the phase is spatially constant, the amplitude changes with time according to the oscillating amplitude of the wave at the interface between normal and n-near-zero material. The results in Ref. [Zio04] were obtained for a Drude type permittivity and permeability with identical frequency dependence and identical losses Γ

$$\begin{aligned}\varepsilon(\omega) &= \varepsilon_0\left[1-\frac{\omega_p^2}{\omega(\omega-i\Gamma)}\right]\\ \mu(\omega) &= \mu_0\left[1-\frac{\omega_p^2}{\omega(\omega-i\Gamma)}\right].\end{aligned} \tag{2.57}$$

Unfortunately, the above assumptions of identical frequency dependence - implying an impedance match to vacuum -, as well as identical losses are difficult or even impossible to implement. On the other hand, in order to obtain an index of refraction that is zero, only one of the two material parameters needs to be zero.

This leads to the second case, namely ε-near-zero (ENZ) materials. Since $\mu \neq 0$, the rotation of the electric field (Eq. (2.55)) does not vanish. However, assuming without loss of generality $\vec{H}(\vec{r}) = H_z(\vec{r})\hat{e}_z$, the second Maxwell equation (cf. Eqs. (2.43)-(2.46)) can be transformed to

$$\frac{\nabla H_z(\vec{r}) \times \hat{e}_z}{-i\omega\varepsilon} = \vec{E}(\vec{r}). \tag{2.58}$$

Since $\vec{E}(\vec{r})$ has to be finite, $\nabla H_z(\vec{r})$ must be zero, which, in turn means, that H_z and hence the phase of the full electromagnetic wave is constant inside a ENZ material [SE06]. Using this result, the authors of Ref. [SE06] theoretically predicted that light can tunnel through an arbitrarily shaped 2D waveguide section consisting of an ENZ material as long as its cross-section is small compared to the vacuum wavelength and under the assumption, that light enters the ENZ material perpendicular to its surface. This theoretical prediction was later experimentally verified [Edw+08]. Other potential applications of ENZ materials include their use as wave front transformers [Zio04; Al07] and the development of highly directive sources [Eno+02]. The opposite effect, namely all-angle collimation was predicted by Feng [Fen12] for the case, when p-polarized light enters a lossy ENZ medium.

Although less research was done on the topic of μ-near-zero (MNZ) materials, this last effect was also found for the case, when s-polarized light enters a MNZ medium [FN13]. For a MNZ material, the same considerations are applicable as for the ENZ medium, except that now $\nabla \times \vec{E}(\vec{r})$ has to be used and the polarization has to be adjusted to s-polarised light, accordingly.

2.4 Josephson Metamaterials

2.4.1 The rf-SQUID as Meta-Atom

In Sec. 2.3 of this chapter, it was explained how to build a magnetic meta-atom using a split ring resonator. It was shown why the resonant behavior of the SRR is crucial for achieving values for μ_r that are less than unity or even negative. On the other hand, it was pointed out that due to the resonant nature of such a meta-atom, the frequency range where μ_r deviates from unity is limited.

Considering the properties of the rf-SQUID as presented in Sec. 2.1.3, it seems to be obvious to replace the SRR with the rf-SQUID as suggested previously in Refs. [DCL06; LT07]. Thereby the strong frequency limitation of the SRR is circumvented due to the tunable resonance frequency (Eq. (2.17)) of the SQUID.

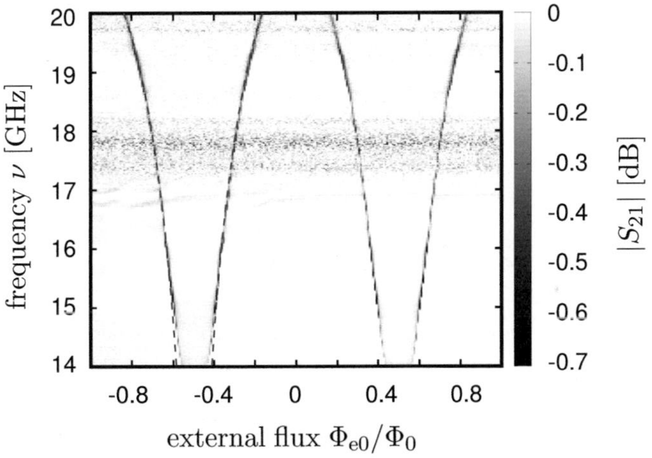

Fig. 2.11: Measured transmission magnitude through a coplanar waveguide containing one rf-SQUID in dependence of frequency and magnetic flux. The calculated magnetic flux dependent resonance curve (Eq. 2.17) is shown by a dashed blue line. Figure courtesy of Ref. [Jun+13].

In Ref. [Jun+13], it was shown that for small driving power, the single rf-SQUID indeed behaves as predicted. The SQUID was placed into a coplanar waveguide (CPW), oriented such that the inductive coupling to the magnetic field component was maximized. The transmission magnitude and phase through this CPW was recorded over a wide frequency range in dependence of an additional, externally applied, constant magnetic field. Whenever the microwave signal was in resonance with the rf-SQUID, a decreased transmission was observed as shown in Fig. 2.11. The measured resonance curve agrees well with the curve calculated from the SQUID parameters and is periodic in Φ_0.

It should be noted, that an additional parallel plate shunt capacitance is included in the SQUID in parallel to the Josephson junction. This is necessary to decrease the resonance frequency, since the measurement setup does not support frequencies above 20 GHz. Nonetheless, the zero flux resonance frequency of this particular SQUID lies still above the 20 GHz and the part of the resonance curve continuing to higher frequencies cannot be shown.

2.4.2 Transmission Line Setup

In the common understanding, a material extends in three dimensions, especially in the context of refraction. On the other hand, the construction of a material out of artificial meta-atoms is easier, the lower the dimensionality. For example, the first experiment demonstrating a negative index of refraction involved a quasi two dimensional metamaterial. The thickness of the medium was only three meta-atomic layers. A 2D waveguide defined the direction of propagation, which lay in the quasi 2D plane [SSS01]. Since in an isotropic and homogeneous medium, incoming and refracted beam lie in one plane, such a 2D approach is a valid design for a metamaterial. When extending the concept of metamaterials to media that are built using superconducting meta-atoms, the low temperature setup and potential application in combination with circuit quantum electrodynamics suggests the implementation of a 1D metamaterial.

For this work, a coplanar design is chosen, where one dimensional arrays of, for example, rf-SQUIDs are placed in the two gaps of a transmission line as schematically shown in Fig. 2.12(a). This setup makes it possible to use preexisting electronic microwave devices and a technically fully developed experimental setup. Additionally, a good coupling between transmission line, i.e. microwave signal, and rf-SQUID is achieved which is relevant especially for first proof-of-principle measurements on this novel metamaterial. Further advantages include a simple, quasi TEM mode structure in the coplanar waveguide (cf. Sec. 2.2.1). Since other, non QTEM modes are not supported in the CPW (unlike e.g. rectangular waveguides), this leads to a favorable magnetic and electric field orientation in the gap. Additionally, the CPW design offers the possibility to use the central conductor also for a constant magnetic field bias.

For this last purpose, a constant bias current I_b is applied to the central conductor in addition to the microwave signal.

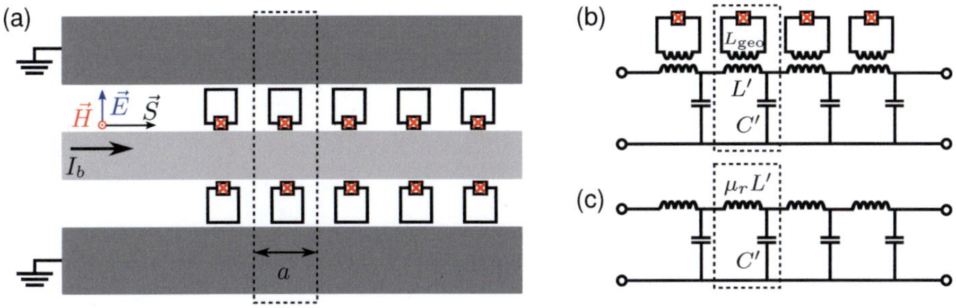

Fig. 2.12: (a) Sketch of a SQUID loaded coplanar waveguide as seen in top view. The ground plane is illustrated in dark, the central conductor in light gray. The dashed rectangle indicates one unit cell of length a. (b) Lumped-element equivalent circuit of the unloaded transmission line ($L' = l'a$, $C' = c'a$) and the inductive coupling of the rf-SQUIDs. (c) The effect of the rf-SQUIDs is included into the effective magnetic permeability of the transmission line.

Although this work will focus solely on the 1D coplanar waveguide setup, it should be noted that rf-SQUIDs have been successfully employed as magnetic meta-atoms in a 2D geometry inside a rectangular waveguide [Tre+13]. As expected, the coupling and hence the signal of the resonance of the rf-SQUIDs was much weaker in this setup.

retrieval of the magnetic permeability

Already before the invention of metamaterials, the question of how to extract the material parameters from reflection and transmission coefficient was answered. Nicolson and Ross [NR70] and Weir [Wei74] considered a $50\,\Omega$ transmission line that contained a dielectric material of thickness d with material parameters μ and ε. They showed, that by knowledge of the scattering matrix S at predefined reference planes, these parameters can be determined. Therefore, they used the reflection coefficient Γ and the transmission coefficient z at the interface of and through the dielectric, respectively. The transmission coefficient $z = \exp(-i\omega\sqrt{\mu\varepsilon}d)$ contains the material parameters directly.

The approach developed in this thesis differs from the above procedure only in the means used to connect S-matrix and material parameters. Employing

z and Γ requires exact knowledge of the position of the reference planes, i.e. their distance to the surfaces of the dielectric and the length of the dielectric. This is not possible in our measurements, since the setup does not allow a full calibration. Instead, the ABCD matrix of one unit cell is used. This effectively places the reference planes directly in front of and behind the sample.

Since the capacitive coupling of the SQUID to the microwave signal is negligible compared to the inductive coupling, the SQUIDs only couple to the inductance of the line. This effect is projected onto a relative magnetic permeability μ_r which in turn modifies the inductance of the transmission line as illustrated in Fig. 2.12(b) and (c). Note that this approach is similar as the one presented in Sec. 2.3.3 for transmission line metamaterials. The ABCD matrix of one unit cell can then be written as

$$T_{\mathrm{m}} = \begin{pmatrix} \frac{Z_{\mathrm{L}}^{\mu}}{Z_{\mathrm{C}}} + 1 & Z_{\mathrm{L}}^{\mu} \\ \frac{1}{Z_{\mathrm{C}}} & 1 \end{pmatrix} . \tag{2.59}$$

Here, $Z_{\mathrm{L}}^{\mu} = i\omega\mu_r l' a$ is the impedance of the inductance and $Z_{\mathrm{C}} = 1/(i\omega c' a)$ is the impedance of the capacitance for a unit cell of length a. As a next step, the ABCD matrix of N unit cells is found by cascading the matrix T_{m} N times (cf. Sec. 2.2.2). Therefore, it is useful to find the diagonalizing and diagonalized matrices. Hence, the eigenvalues

$$\begin{aligned} l_1 &= \frac{Z_{\mathrm{L}}^{\mu} + 2\,Z_{\mathrm{C}} - \sqrt{(Z_{\mathrm{L}}^{\mu})^2 + 4\,Z_{\mathrm{L}}^{\mu} Z_{\mathrm{C}}}}{2\,Z_{\mathrm{C}}}, \\ l_2 &= \frac{Z_{\mathrm{L}}^{\mu} + 2\,Z_{\mathrm{C}} + \sqrt{(Z_{\mathrm{L}}^{\mu})^2 + 4\,Z_{\mathrm{L}}^{\mu} Z_{\mathrm{C}}}}{2\,Z_{\mathrm{C}}} \end{aligned} \tag{2.60}$$

and eigenvectors

$$e_1 = \left(1, -\frac{l_2 - 1}{Z_{\mathrm{L}}^{\mu}}\right)^T, \qquad e_2 = \left(1, -\frac{l_1 - 1}{Z_{\mathrm{L}}^{\mu}}\right)^T \tag{2.61}$$

of the ABCD matrix T_{m} need to be found. Then T_{m} can be rewritten as

$$T_{\mathrm{m}} = D \cdot \begin{pmatrix} l_1 & 0 \\ 0 & l_2 \end{pmatrix} \cdot D^{-1} \tag{2.62}$$

$$\text{with } D = \left(e_1, e_2 \right) . \tag{2.63}$$

This ABCD matrix T_{m} then is cascaded N times in order to obtain the full ABCD matrix $\mathscr{T}_{\mathrm{m}}$ of the metamaterial:

$$\mathscr{T}_{\mathrm{m}} = T_{\mathrm{m}}^N = D \cdot \begin{pmatrix} l_1^N & 0 \\ 0 & l_2^N \end{pmatrix} \cdot D^{-1}. \tag{2.64}$$

As mentioned above, there are several approaches, about how to extract the complex material parameters ε and μ from the scattering matrix, using S_{11} and S_{21} [NR70; Wei74; BJVK90]. Since the deviation of the electric permittivity from unity is negligible for this metamaterial, we are looking only to find the magnetic permeability. By eliminating one complex unknown quantity, i.e. ε_r, using only the transmission coefficient S_{21} is sufficient to determine μ_r.

In order to find the relative magnetic permeability, the elements of the ABCD matrix $\mathscr{T}_{\mathrm{m}}$ of the full metamaterial are expressed in two different ways. First, in dependence of the S-matrix elements [Poz05], it is given by

$$\mathscr{T}_{\mathrm{m,S}} = \begin{pmatrix} -\frac{(S_{11}+1)S_{22}-S_{12}S_{21}-S_{11}-1}{2\,S_{21}} & \frac{((S_{11}+1)S_{22}-S_{12}S_{21}+S_{11}+1)Z_0}{2\,S_{21}} \\ \frac{(S_{11}-1)S_{22}-S_{12}S_{21}-S_{11}+1}{2\,S_{21}Z_0} & -\frac{(S_{11}-1)S_{22}-S_{12}S_{21}+S_{11}-1}{2\,S_{21}} \end{pmatrix} \tag{2.65}$$

Second, the matrix $\mathscr{T}_{\mathrm{m}}$ is written in terms of eigenvalues l_1 and l_2 of the ABCD matrix of the unit cell and its impedance parameters Z_L and Z_C, i.e. substituting Eqs. (2.60), (2.61) and (2.63) into Eq. (2.64):

$$\mathscr{T}_{\mathrm{m,\,eig}} = \begin{pmatrix} -\frac{(l_2-1)l_2^N-(l_1-1)l_1^N}{l_1-l_2} & \frac{\left(l_1^N-l_2^N\right)Z_{\mathrm{L}}^{\mu}}{l_1-l_2} \\ -\frac{(l_1-1)l_1^N l_2-(l_1-1)l_1^N-((l_1-1)l_2-l_1+1)l_2^N}{(l_1-l_2)Z_{\mathrm{L}}^{\mu}} & \frac{(l_1-1)l_2^N-l_1^N l_2+l_1^N}{l_1-l_2} \end{pmatrix} \tag{2.66}$$

By equating both expressions for $\mathscr{T}_{\mathrm{m}}$, a coupled system of four nonlinear equations is obtained. The corresponding four unknown (complex) variables are S_{11}, S_{12}, S_{22}, and μ_r. The latter is hidden in Z_{L}^{μ}, which in turn is included in l_1 and l_2. S_{21} is the measured quantity and all other parameters (e.g. L' and C') are known from either design considerations or simulations.

Although the approach uses the ABCD matrix of the system instead of transmission (z) and reflection (Γ) coefficient at the interface (cf. Refs. [NR70; Wei74; BJVK90]), similar problems arise. Reference planes directly in front of and behind the sample are chosen, which still requires a suitable calibration as detailed later in Sec. 3.2.4. Also, there can be up to N different solutions when extracting the N-th root during the retrieval process. In order to test whether the correct root is found, the measured phase delay is compared to the calculated phase delay, as suggested also by Weir [Wei74]. For the calculated delay, the obtained result of μ_r is used. Another option involves testing of energy conservation by making sure that the determinant of the scattering matrix S is less than unity.

The task of solving the system of equations and testing the obtained solution is left to a computer. In order to facilitate the search for the correct root, suitable start values have to be supplied to the solver. Far away from resonance, i.e. for frequencies where $\mu_r \approx 1$, the transmission line can be considered to behave like an unloaded transmission line matched to $50\,\Omega$. Hence, acceptable start values would be $Z_L = i\omega l_r a$, $S_{21} = S_{12} = 1$ and $S_{11} = S_{22} = 0$.

As the frequency approaches the resonance frequency, μ_r changes continuously. It is thus reasonable to assume, that the solution of the previous frequency value is a good start value for the next frequency point, provided that the step between two frequency points is sufficiently small. The latter requirement is also necessary for the phase delay test. Hence, as first start values, parameters corresponding to an unloaded transmission line are chosen at low frequency. The solution of this first result is used as new start value for the next frequency value and so forth until the highest frequency is reached.

This procedure will be applied to the calibrated transmission data measured from the magnetic metamaterial in Chapter 4.

2.4.3 A Metamaterial made from Tunable, Electric Meta-Atoms

In Sec. 2.3.2, thin metallic wires were introduced as one way of constructing an electrically interacting metamaterial. This approach is not feasible in the chosen CPW design discussed above. Adding thin superconducting wires in the two gaps oriented parallel to the electric field component comes with the problem that the coupling between $\vec{E}$-field and rods is negligible. A strong coupling could be achieved by connecting the wires galvanically to central conductor and ground. In this case, the CPW is shortened periodically to ground. Note, that a similar setup is often used in the case of a rectangular or parallel plate waveguides. In this case, the wires are also connected to the conducting plates of the waveguide, but the mode structure and the field distribution in these waveguides is not comparable to the CPW.

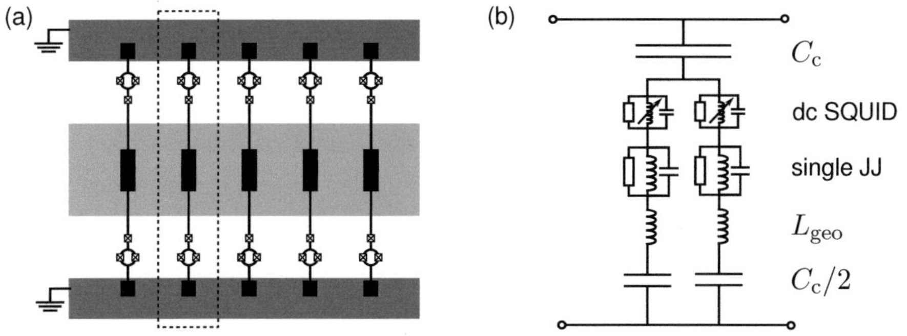

Fig. 2.13: (a) Sketch of a CPW loaded with capacitively coupled rods. The ground planes are illustrated in dark, the central conductor in light gray. The black rectangles symbolize parallel plate capacitors between rods and waveguide. Crosses in boxes are Josephson junctions. The unit cell is indicated by the dashed rectangle. (b) Electric circuit equivalent of one rod structure coupling to the transmission line.

To circumvent these problems, an electrically coupling meta-atom was developed that couples capacitively to the CPW. Since it also comprises an inductance, such a structure is of resonant nature. As for SRR and rf-SQUID, the interesting behavior is in the frequency range around and above the resonance frequency, where its susceptibility deviates from zero.

A sketch of a CPW containing five of these structure coupled to the transmission line is shown in Fig. 2.13(a).

A dc-SQUID is installed in the wire to create a magnetic field tunable inductance (cf. Sec. 2.1.3). Since its loop inductance is small compared to the Josephson inductance, it behaves like a single junction but with magnetic field tunable critical current. Additionally, it decreases the resonance frequency of the rod structure. Since this decrease is not sufficient to reach a resonance frequency accessible by the measurement setup, an additional single Josephson junction is included in series to the dc-SQUID. Junctions are symbolized as black crosses in Fig. 2.13(a). The critical current of the junctions is optimized, so that resonance frequency of the rods is within the range of tunability of the magnetic meta-atom. Only then a combination of both kinds of meta-atoms may yield an overlapping range of negative μ and ε.

The electric circuit equivalent of one rod is shown in Fig. 2.13(b). The dc-SQUID is depicted as a single Josephson junction with a tunable critical current, i.e. Josephson inductance. The maximum critical current of this effective junction is twice the critical current of the single junction, hence its zero magnetic field inductance is only half the inductance of the single junction. The same is true for the resistance, while the capacitance is doubled compared to the single junction.

In order to understand the transmission properties of such a transmission line loaded with electrically coupling elements, the ABCD matrix of one unit cell is found

$$T_\mathrm{e} = \begin{pmatrix} \frac{Z_\mathrm{L}}{Z_\mathrm{C}} + 1 & Z_\mathrm{L} \\ \frac{1}{Z_\mathrm{C}} & 1 \end{pmatrix} \begin{pmatrix} 1 & 0 \\ \frac{1}{Z_\mathrm{rod}} & 1 \end{pmatrix} \begin{pmatrix} \frac{Z_\mathrm{L}}{Z_\mathrm{C}} + 1 & Z_\mathrm{L} \\ \frac{1}{Z_\mathrm{C}} & 1 \end{pmatrix}. \tag{2.67}$$

Here, the first and the last matrix combined describe the effect of the unloaded transmission line and are defined for half the length of a unit cell by $Z_\mathrm{L} = i\omega L'/2$ and $Z_\mathrm{C} = 2/(i\omega C')$. L' and C' are the characteristic inductance and impedance of the unloaded line per unit length a. The symmetric arrangement insures a symmetric and thus reciprocal unit cell. However, it should be noted that the use of a symmetric cell is no longer crucial, if a larger number of unit cells is cascaded. The matrix in the center is the ABCD matrix of only the rod.

The impedance of the rod Z_rod is a combination of the impedances of the different elements (cf. Fig. 2.13(b)) and given by

$$Z_\mathrm{rod} = \frac{2}{i\omega C_\mathrm{c}} + \frac{1}{2}(Z_\mathrm{geo} + Z_\mathrm{JJ} + Z_\mathrm{SQ}(\Phi_\mathrm{e0})), \tag{2.68}$$

with

$$Z_\mathrm{geo} = i\omega L_\mathrm{geo}, \tag{2.69}$$

$$Z_\mathrm{JJ} = \left(\frac{1}{i\omega L_\mathrm{j0}} + \frac{1}{R} + i\omega C_\mathrm{in}\right)^{-1}, \tag{2.70}$$

$$Z_\mathrm{SQ}(\Phi_\mathrm{e0}) = \left(\frac{2\cos(2\pi\Phi_\mathrm{e0}/\Phi_0)}{i\omega L_\mathrm{j0}} + \frac{2}{R} + 2i\omega C_\mathrm{in}\right)^{-1}. \tag{2.71}$$

The definitions of L_geo and C_c are given in Fig. 2.13(b). L_j0 is the Josephson inductance of the single junction, R and C_in are its normal resistance and intrinsic capacitance. The first term of impedance of the dc-SQUID Z_SQ takes into account the flux tunability of the effective critical current of the SQUID.

Due to the combination of inductive, resistive and capacitive elements in the rod, the impedance of the rod is no longer purely imaginary and, additionally, it is of resonant nature.

periodically loaded line

By cascading the ABCD matrix T_e of the unit cell (Eq. (2.67)) N times, the ABCD matrix $\mathscr{T}_e$ of the complete, loaded transmission line is found. From $\mathscr{T}_e$, the transmission coefficient S_{21} can be determined. Its frequency dependence at zero magnetic flux for a transmission line loaded with 27 rod structures is shown in Fig. 2.14(a).
Next, using Eq. (2.37), the dispersion relation is calculated by combining phase advance and ABCD matrix of one unit cell as explained in Sec. 2.2.3. Figure 2.14(b) shows the real part β (blue) and the imaginary part α (red) of the resulting propagation constant. Since the rod impedance is of resonant nature and lossy, there is a frequency range around the resonance frequency where β and α are simultaneously nonzero. The resonance is also the reason why the dispersion relation is not unique in this frequency range. However, as soon as α is nonzero, the corresponding wave is strongly attenuated. Comparing Fig. 2.14(a) and (b) shows how the onset of the stop band ($\beta \approx 0$) coincides with the onset of the resonance at about 10.2 GHz.

Furthermore, below the resonance frequency, i.e. below 10 GHz, good transmission is observed only at discrete frequencies. This is due to the finite length of the periodic loading, as explained in Sec. 2.2.3. In order to illustrate this effect, Fig. 2.14(c) shows the wavelength in units of unit cells in dependence of frequency. The solid line depicts the continuous case which is valid for an infinite transmission line. It is calculated from the imaginary part of the propagation constant, β. Once $\beta \approx 0$ (it never becomes truly zero) above 10.2 GHz, the wavelength jumps to large values at that frequency. The red circles mark the frequencies of good transmission. At these frequencies integer numbers of half wavelengths are equal to the length of the loaded line (27 unit cells). The first circles are at $\lambda = 54a$, $27a$, and $13.5a$. This is an effect which is also known from transmission line resonators. Good transmission in such a resonator is observed when it is at resonance, i.e. whenever integer numbers of half wavelength fit into the length of the resonator. The same principle applies for the rod loaded transmission line since there is an discontinuity in impedance at both ends.

Unlike the rf-SQUIDs, whose interaction with the transmission line is negligible as long as they are not at resonance, the 1D rod array affects the transmission strongly at any frequency. This is clearly visible when comparing the frequency dependence of the real part of the Bloch impedance (cf. Eq. (2.40)) as shown in blue Fig. 2.14(d) to the standard 50 Ω.

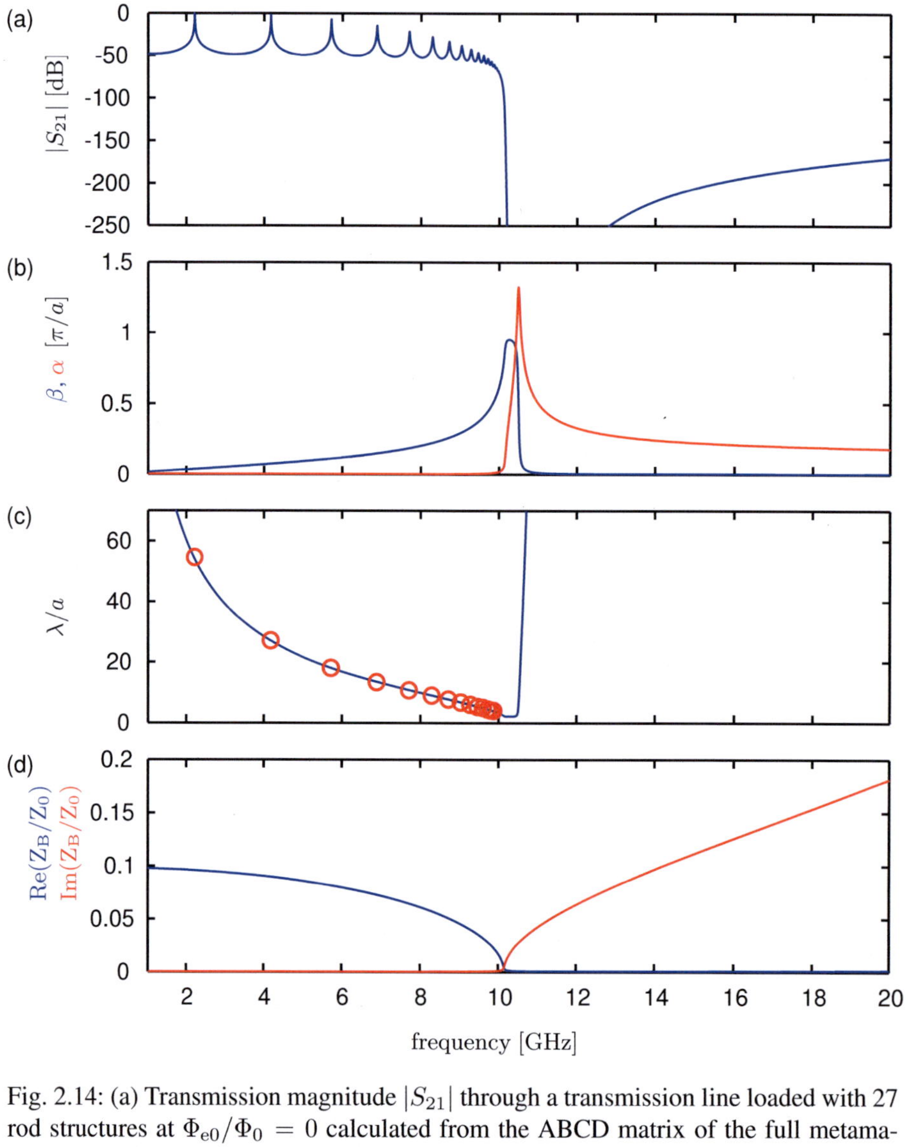

Fig. 2.14: (a) Transmission magnitude $|S_{21}|$ through a transmission line loaded with 27 rod structures at $\Phi_{e0}/\Phi_0 = 0$ calculated from the ABCD matrix of the full metamaterial. (b) Corresponding real (blue) and imaginary (red) frequency dependence of the wave vector $\beta + i\alpha$. (c) Frequency dependent wavelength λ in units of unit cell lengths. The blue line depicts the continuous case, while the red circles indicate the discrete frequencies and corresponding wavelengths of good transmission. (d) Real (blue) and imaginary (red) part of the Bloch impedance Z_{B} of one unit cell.

The real part of the impedance is less than 10 % of the impedance of the unloaded line. At resonance ($\nu \approx 10.2$ GHz), the imaginary part of the Bloch impedance starts to deviate from zero and no transmission is possible anymore. Considering the small Bloch impedance (compared to 50 Ω) and corresponding short wavelength, the question needs to be asked, to which extend the definition of effective parameters and an index of refraction makes sense.

As explained in Ref. [Smi+05], in a periodic system an effective index of refraction can be defined even for $\lambda \sim a$, since the dispersion relation is known from only one unit cell. This index of refraction can be related to frequency dependent "material parameters" although the concept of homogenization does not apply anymore. Similarly, the electric permittivity of the rod array can be determined from the dispersion relation under the assumption that the magnetic permeability of the rod array is $\mu_r = 1$. In the stop band, the value of the permittivity is strongly negative, as expected from the large impedance mismatch.

tunability

In the frequency range shown in Fig. 2.14, only the first pass band and the stop band is observable. The second pass band cannot be observed since it starts at approximately 40 GHz at zero flux.

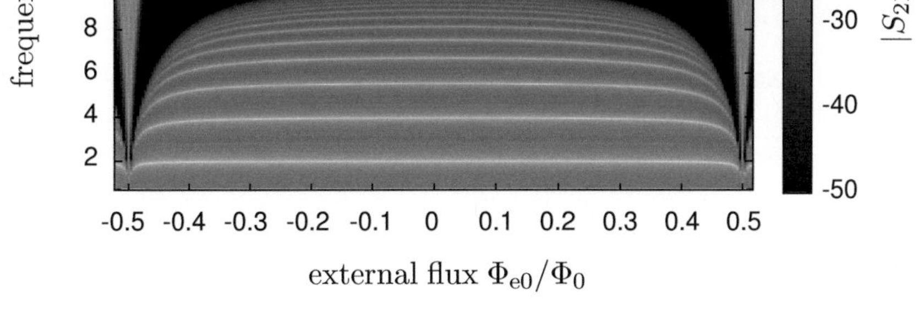

Fig. 2.15: Flux and frequency dependence of the transmission magnitude through a transmission line loaded with 27 rod structures.

Figure 2.15 shows the calculated transmission magnitude in gray scale for an external flux Φ_{e0} that varies between $\pm 0.52\Phi_0$. Since the inductance of the dc-SQUID approaches infinity, the onset of the stop band and also of the second pass band approaches zero. Therefore, the onset of the second pass band becomes visible as Φ_{e0} approaches $\pm 0.5\,\Phi_0$.

2.4.4 Tunable Composite Metamaterial

By combining both tunable rf-SQUIDs and tunable rods in the CPW, a one dimensional tunable metamaterial is created. In the stop band of the rod array, only decaying waves exists due to the negative permittivity. However, at frequencies for which the magnetic permeability of the magnetic SQUID submaterial becomes also negative, wave propagation is allowed. This is illustrated in Fig. 2.16, which depicts the calculated transmission magnitude of the bare rod (black), bare SQUID (blue) and composite (red) medium. The transmission coefficient is obtained in all three cases by cascading the respective unit cell N times, where N is again the number of unit cells.

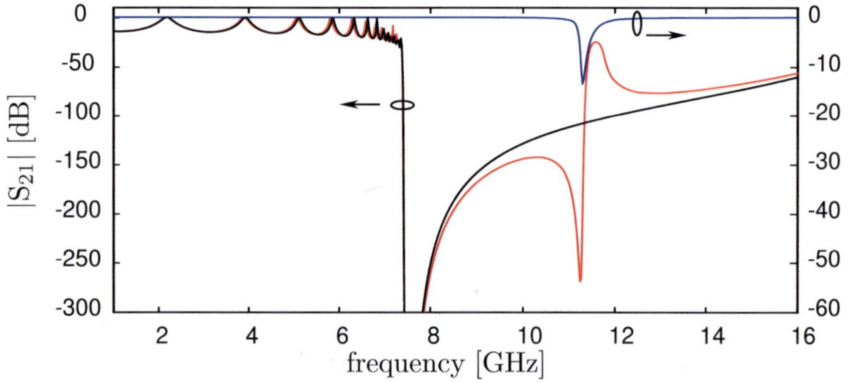

Fig. 2.16: Transmission magnitude through a medium containing 27 rod structures (black) at a flux $\Phi_{e0} = 0.44\Phi_0^{\rm rod}$, 27×2 rf-SQUIDs (blue) at a flux $\Phi_{e0} = 0.37\Phi_0^{\rm SQUID}$ and 27 composite unit cells (red). For better visibility, the transmission magnitude through the SQUID metamaterial is multiplied with a factor 5.

The ABCD matrix of the rf-SQUID loaded transmission line unit cell is given by Eq. (2.59). It contains the magnetic permeability which needs to be obtained from simulation as follows. First, the single meta-atom susceptibility as given in Eq. (2.51) is calculated from the time average of the simulated steady state solution of the differential equation describing the dynamics of the rf-SQUID (cf. Eq. (2.16)). Second, the filling factor $\tilde{F}$ (cf. Sec. 2.3.2), necessary to determine the relative magnetic permeability, has to be found. The relevant volume is approximated by a torus shape with its axis oriented along and placed in the center of the central conductor of the CPW. The gap of the coplanar waveguide defines inner and outer radius of the torus and the length of the unit cell a defines its width. The magnetic moment in this volume is given by the ratio of area and loop inductance of the SQUID. More details concerning the filling factor $\tilde{F}$ are given in Ref. [Jun+13].

Combining filling factor and single meta-atom susceptibility yields the magnetic permeability. It is then included into Z_{L}^{μ} which contains also the contribution of the unloaded line (cf. Eq. (2.59)).

The ABCD matrix for the electrical component is given in Eq. (2.67). This equation is also used to find the ABCD matrix of the unit cell of the composite medium. Therefore, the ABCD matrix of the unloaded transmission line in Eq. (2.67) is replaced by the ABCD matrix of a transmission line containing one rf-SQUID (cf. Eq. (2.59)). The combined matrix of one unit cell is then

$$T_{\mathrm{comp}} = \begin{pmatrix} \frac{Z_{\mathrm{L}}^{\mu}}{Z_{\mathrm{C}}} + 1 & Z_{\mathrm{L}}^{\mu} \\ \frac{1}{Z_{\mathrm{C}}} & 1 \end{pmatrix} \begin{pmatrix} 1 & 0 \\ \frac{1}{Z_{\mathrm{rod}}} & 1 \end{pmatrix} . \tag{2.72}$$

This ABCD matrix is not symmetric which allows us to use the ABCD matrix of SQUID loaded unit cell as defined in Eq. (2.59). However, as pointed out above, a symmetric definition is not necessary since only the ABCD matrix of N unit cells is of interest. However, if necessary, a symmetric unit cell could be defined easily.

Due to the different areas of the rf-SQUID and the small dc-SQUID in the rod structure, their periodicity in external flux is different. One flux quantum in the small dc-SQUID corresponds to approximately 30 flux quanta in the rf-SQUID. This means, that the position of the resonance of the magnetic sub-material can be chosen almost arbitrarily with respect to the resonance of the electric sub-material.

3 Experimental Setup and Procedures

This chapter starts by presenting the different samples used in this work including a brief description of the fabrication process. Next, it introduces the experimental principles, for example the used setup and details on the measurement calibration. Additionally, it demonstrates how the setup has to be optimized in order to overcome one of the biggest obstacles, namely the effect of stray magnetic fields which destroys the common resonance of the magnetic meta-atoms.

3.1 Sample Design and Fabrication

The metamaterials under consideration in this thesis are all implemented in a 1D waveguide geometry as sketched in Fig. 2.12 in the previous chapter. The parameters, i.e. critical current, inductances and capacitances, of both electric and magnetic meta-atoms define the range of tunability of the respective resonance frequency and have to be chosen accordingly.

3.1.1 Fabrication

All samples discussed in this work were produced at the Kotel'nikov Institute of Radio Engineering and Electronics (IREE RAS) in Moscow, Russia. In each fabrication run, an array of 16 samples is fabricated on a 1 inch silicon wafer using a Nb/Al0_x/Nb trilayer process. After fabrication, the wafer is diced into 4×4 mm^2 chips, each containing one or two coplanar waveguides with the respective structures.

The fabrication process starts by depositing the Nb/AlO_x/Nb trilayer. It consists of the Nb base electrode with a thickness of 300 nm, a thin (∼1 nm) layer of AlO_x, which later serves as the isolating layer of the Josephson junction and another Nb layer on top. After deposition, the shape of the base electrode is patterned into the trilayer. Subsequently, the top Nb layer is removed in the area where the parallel plate shunt capacitor will be located. This area is anodized with 40 V to create a Nb_2O_5 layer with a thickness of 30 nm (in addition to the 1 nm AlO_x of the trilayer). Some samples use normal metal ground plates made from palladium (Pd) instead of superconducting ones. This resistive layer is deposited after the anodization process.

In the next step, the area of Josephson junctions and vias is defined by first removing the top Nb layer everywhere except where junctions and vias are supposed to be later (for the capacitor it was already removed in the previous step). Then, the uncovered area, i.e., the area of the base electrode with exception of the junctions, the vias and the parallel plate capacitor, is lightly anodized (10V). This forms an insulating, mechanically stabilizing wall around the junctions and vias. Finally, SiO_2 is deposited on the same area.

Before the Nb layer, which will form the top electrode of the structure, is deposited, the top layer and the AlO_x layer of the trilayer has to be etched away wherever direct contact between top and bottom electrode is needed, i.e. at the vias. In order to obtain good contact between the top electrode and the top of the junction or the bottom electrode, an additional etching process is used just before the top electrode Nb deposition. As a last step, a gold layer is deposited at the contact pads, in order to provide bonding pads with good bonding properties.

3.1.2 Samples

The experimental results that will be presented in this thesis are obtained from three different samples. Each sample contains different meta-atoms, either only magnetically coupling rf-SQUIDs (M) or only electrically coupling rods (E). Except where stated otherwise, the ground planes are made from the normal metal Pd instead of superconducting Nb. The CPW parameters are adjusted to the size of the respective meta-atom and chosen such that the unloaded transmission line is matched to $Z_0 = 50\,\Omega$. They are

w	width of the central conductor in [μm]
g	width of the gap in [μm]
l'	inductance per unit length in [nH/m]
c'	capacitance per unit length in in [pF/m].

Additionally, the number of unit cells N and the length of the unit cell a define the transmission line metamaterial. The parameters of the transmission lines are given in Tab. 3.2 at the end of this section on p. 62.

magnetic meta-atom

As magnetic meta-atom two different variations of the rf-SQUID are used. The relevant parameters that define the flux and the frequency dependence are

I_c	the critical current of the junction in [μA]
L_{j0}	the Josephson inductance at zero magnetic field in [pH]
L_{geo}	the geometric inductance of the SQUID loop in [pH]
β_L	the ratio of $L_{\mathrm{geo}}/L_{\mathrm{j0}}$
C	capacitance of the additional parallel plate shunt capacitor together with the intrinsic capacitance of the junction in [pF]
R	normal resistance of the junction in [Ω]
M	mutual inductance between transmission line and SQUID in [pH].

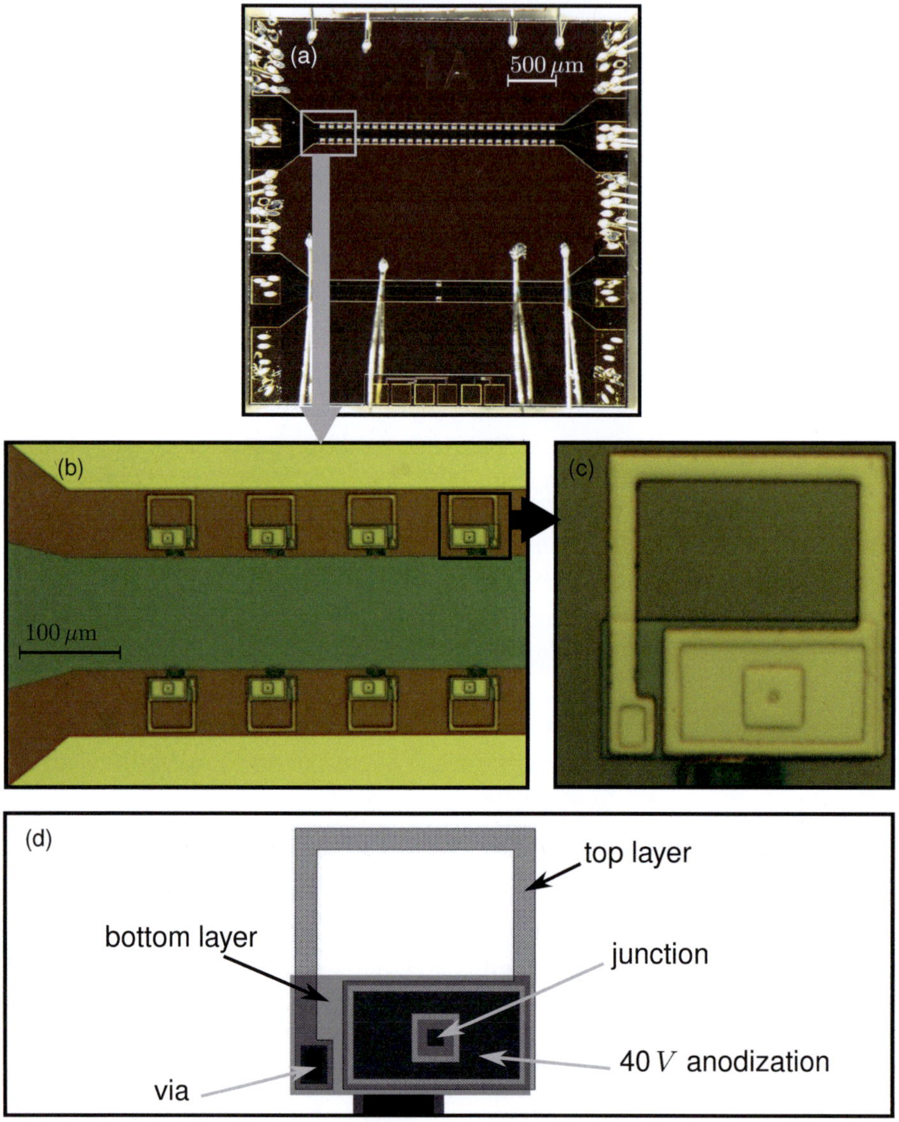

Fig. 3.1: (a) Optical micrograph of the full chip of sample M1. The structures at the bottom of the picture are test junctions. (b) Zoom to part of the CPW loaded with rf-SQUIDs. The central conductor is shown in green, the ground plates in yellow. (c) Optical micrograph of the individual SQUID used in sample M1. (d) Picture of the design of one rf-SQUID used in sample M1. The relevant details and layers are pointed out in the figure.

As already mentioned, the additional parallel plate capacitor is necessary to decrease the resonance frequency of the SQUID to a frequency range, that is accessible with the available measurement setup. The parameters of the SQUIDs used in sample M1 and M2 are given in Tab. 3.3 on p. 62.

An optical micrograph of sample M1 (U02_32 chip1A) is depicted in Fig. 3.1(a). The picture was taken using a optical stereo microscope. The chip contains two transmission lines, one incorporating two (bottom) and one incorporating 54 rf-SQUID meta-atoms (top). The 54 SQUIDs ($N = 27$) in the latter are evenly spaced with a unit cell length $a = 92\,\mu$m and distributed symmetrically in the two gaps. Only results obtained with this transmission line are shown later, the transmission line containing only two SQUIDs is not considered. The white lines connecting to the edges of the chip and to the ground plane in the center are bond wires. Part of the waveguide is shown in Fig. 3.1(b), 2×4 SQUIDs are located in the two gaps of the waveguide. This picture is taken with the optical polarization microscope which yields different colors depending on the film thickness and material of the different structures. One single SQUID finally is presented in Fig. 3.1(c). Its different components are best explained using the design shown in Fig. 3.1(d). The black rectangle with the quadratic hole in the center is the parallel plate capacitor, the small dark square in its center the Josephson junction. The small rectangle in the lower left corner is the via connecting top and bottom electrode. The connection between central conductor and SQUID at the bottom of Fig. 3.1(c), is required during the fabrication process. For anodization, the SQUID has to be galvanically connected to a voltage source. This connection is removed later in the fabrication process by etching.

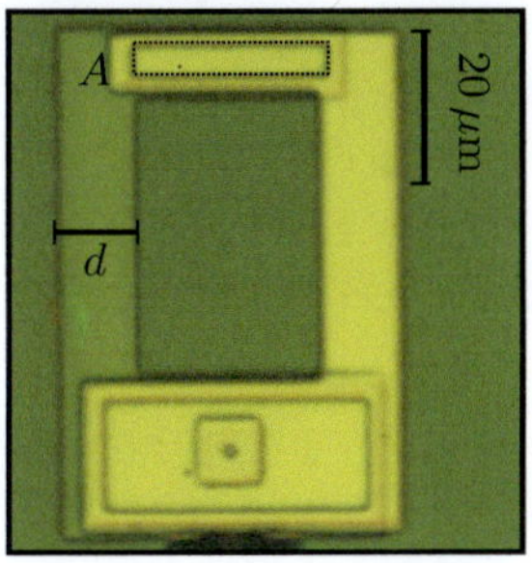

Fig. 3.2: Optical micrograph of the rf-SQUID used in sample M2. The width of the SQUID arms d and the area of the via A is illustrated.

The main difference between sample M2 (U12_22 chip3B) and sample M1 is the width of the superconducting structures and the superconducting ground planes. It is used in Sec. 3.3 to demonstrate how changes to the SQUID design affect the probability to trap Abrikosov vortices, which in turn influences the performance of the metamaterial. In Fig. 3.2, an optical micrograph of the SQUID is shown including the definitions of the structure width d and the area of the via A. The relevance of these quantities will be explained in Sec. 3.3.

electric meta-atom

As electrically coupling elements, the rods introduced in Sec. 2.4.3 are used. They couple capacitively to the central conductor and the ground planes of the waveguide. A small dc-SQUID is used as tunable inductor and an additional Josephson junction is used to decrease its resonance frequency. A sketch of the rod is found in Fig. 3.3. The parameters defining its resonance frequency and frequency dependent behavior are

I_c	the critical current of the junctions in [μA]
L_{j0}	the Josephson inductance of single junction in [pH]
R	normal resistance of the junction in [Ω]
C_{in}	intrinsic capacitance of the junction in [pF]
L_{geo}	the geometric inductance of one arm comprising rod and dc-SQUID loop in [pH]
C_{c}	the capacitance to central conductor which is twice the capacitance to ground in [pF].

Due to its small loop inductance, the dc-SQUID is considered as an effective single junction with a magnetic field tunable critical current. Its effective zero field critical current and its intrinsic capacitance is twice the critical current of the single junction whose parameters are given in the table above. Normal resistance and zero field inductance, on the other hand, are only half the values of the single junction.

Sample E1 (U04_22 chip7B) contains only one transmission line with 27 unit cells of length $a = 100\,\mu$m, i.e. 27 rods. The full chip is shown in Fig. 3.3(a). The picture is taken using the stereo microscope.

In Fig. 3.3(b) part of the transmission line and five rod structures are shown. The pictures was taken with a polarization microscope which causes the color palette. A zoom to only one rod is shown in (c) and a sketch of one rod structure is presented in Fig. 3.3(d). The parameters of sample E1 are also given on p. 62 in Tab. 3.4. The rectangles at either end of the structure and in the center are parallel plate capacitors coupling to the waveguide. A Nb strip of width 4 μm connects the central capacitor to the single Josephson junction, corresponding to a crossed box in Fig. 3.3(d), the single Josephson junction to the dc-SQUID and the dc-SQUID to the ground capacitor. The F-shaped structures in dark blue are necessary during fabrication, since they provide the galvanic contact necessary for anodization. This contact is removed later in the process, however, the effect of the etching process is still visible under the microscope.

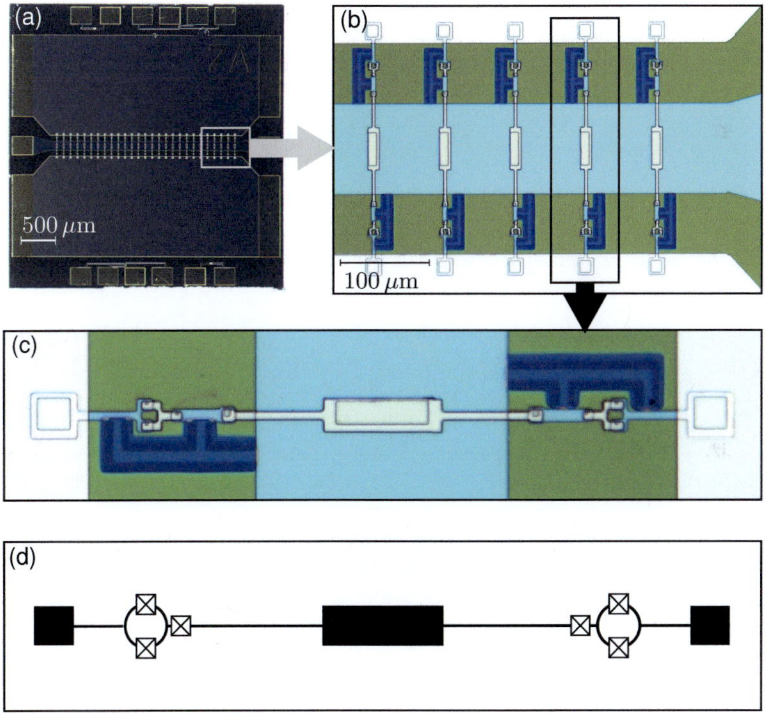

Fig. 3.3: (a) Optical micrograph of the chip, containing the waveguide and test structures (on top and bottom). (b) Optical micrograph of part of the waveguide showing five rods. The light blue, wide strip through the center is the central conductor, the white areas on the top and the bottom are the ground planes. (c) Optical micrograph of one rod structure. (d) Sketch of the rod structure. The black rectangles indicate the parallel plate capacitors, the crossed boxes are again the Josephson junctions.

parameters

	w	g	l'	c'	a	N
M1	100	59	400	160	92	27
M2	130	82	390	155	90	27
E1	124	84	420	164	100	27

Table 3.2: Transmission line parameters. Width of central conductor w, width of gap g, inductance and capacitance per unit length L' and C', length of unit cell a and number of unit cells N. Units on p. 57.

	I_c	L_{j0}	L_{geo}	β_L	C	R	M
M1	1.8	183	83	0.45	2.0	1600	5.7
M2	3.4	97	79	0.80	1.5	890	7.4

Table 3.3: Parameters of the magnetic meta-atom (rf-SQUID). Critical current I_c, geometric inductance L_{geo}, Josephson inductance at zero magnetic field L_{j0}, ratio of the two inductances β_L, mutual inductance between SQUID and transmission line M and capacitance C. Units on p. 57.

	I_c	L_{j0}	R	C_{in}	L_{geo}	C_c
E1	3.2	103	950	0.13	115	3.3

Table 3.4: Parameters of the electric meta-atom (rod). Critical current I_c of the single junction, Josephson inductance at zero magnetic flux L_{j0}, normal resistance R and intrinsic capacitance C_{in} of the single junction, geometric inductance of one arm L_{geo} and capacitance to central conductor C_c which is twice the capacitance to each ground. Units on p. 60.

3.2 Experimental Setup

3.2.1 Mounting of the Sample

As a first step, the chip is glued to a Printed Circuit Board (PCB). A photograph of the PCB including a chip is shown in Fig. 3.4. The PCB itself consists of a ceramic substrate between two copper layers. Into the top copper layer one (or two) coplanar waveguides are structured. They lead from opposite sides of the PCB to the center, where a rectangular hole for the chip is milled. At the other end of each CPW a connector (gold) is soldered that creates the transition from coplanar to coaxial waveguide. The small circular features in the ground plane of the PCB are so-called vias, small holes drilled through the ceramic and filled with conducting material to connect the copper ground plane on top of the PCB with the bottom copper plane.

Fig. 3.4: Printed circuit board with chip.

The chip is connected to its environment by wire bonding the central conductor of the CPW on the chip at both ends to the central conductor of the CPWs on the PCB. Additionally, many wire bonds are used to connect the ground planes of the chip to the ground of the PCB. The bonds, visible as white lines in Fig. 3.4 are Al wires with a 25 μm diameter. While the CPW on chip and PCB are matched to $Z_0 = 50\,\Omega$, the bond connection is not, which may cause reflections.

As second step, the PCB with the chip is installed in one of the two copper sample holders that are depicted in Fig. 3.5. The sample holder stabilizes the connection between connectors on the PCB and the coaxial cables and it protects the sample. Additionally, it serves as body for an coil that produces a magnetic field perpendicular to the area of the chip.

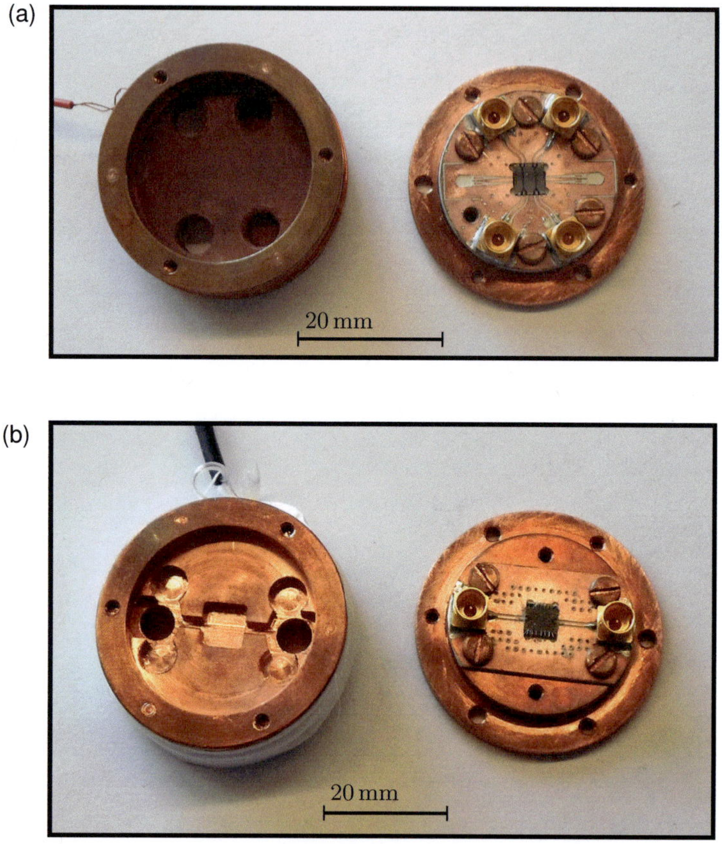

Fig. 3.5: (a) Photograph of sample holder S1, including PCB and chip. It is designed for chips with two transmission lines. (b) Photograph of sample holder S2. When the sample holder is closed, the two halves touch everywhere, except in the area above the chip and above the transmission line on the PCB.

Unfortunately, the copper sample holder is a cavity resonator which, depending on design, can have resonant modes within the frequency range of interest for the measurement. These resonant modes are observed as typically sharp dips in the transmission magnitude at the corresponding frequencies.

The differences in the two sample holders shown in Fig. 3.5 are the volume of the inner cavity and the number and arrangement of transmission lines that are on the corresponding PCB. Sample holder S1, shown in Fig. 3.5(a) is suitable and used for sample M1 and M2. Because of the two transmission lines on the chip, there are four connectors on the PCB and four corresponding holes in the top part of the sample holder. Two of these holes were later closed with a mixture of copper powder and stycast in an attempt to remove parasitic resonances in the sample holder. Although some resonant modes were removed, others occurred and overall, no improvement was detectable.

Sample holder S2 was developed with the goal to shift internal sample holder resonances to higher frequencies. Therefore, the internal cavity was decreased in size as much as possible as can be seen in the left part of Fig. 3.5(b). Only the area is left free which is positioned directly above the chip and above the transmission line on the PCB after closing the sample holder. All the rest touches with the copper of the bottom part (right) in Fig. 3.5. Also for this sample holder, the expected improvement could not be observed.

The reason why no serious improvement was observed is most probably in both cases, that the resonant modes are not located in the empty cavity volume above in the chip, but instead in the dielectric of the chip and the PCB.

3.2.2 Transmission Measurement Setup

Now, that the metamaterial has a connection to the outside world, it can be connected to an experimental setup which allows transmission measurements. The two ports of an Anritsu VectorStar (MS4642A) vector network analyzer (VNA) with a frequency range between 10 MHz and 20 GHz is used as source and detector. It sweeps the frequency at the input and measures the relative complex transmission S_{21}.

Since a logarithmic scale allows a more detailed view of the data at low transmission magnitudes, the data in the next chapter will be shown in logarithmic power scale in units of [dB]. The conversion is given by

$$|S_{21}|\,[\mathrm{dB}] = 20 \cdot \log_{10} |S_{21}|, \tag{3.1}$$

$$|S_{21}|\,[\mathrm{dBV}] = 10 \cdot \log_{10} |S_{21}|. \tag{3.2}$$

The last equation, showing $|S_{21}|$ in logarithmic voltage scale will not be used and is given only of completeness. Note, that only the transmission magnitude is affected, the relative phase δ is the same in all three scales.

A schematic of the whole setup is depicted in Fig. 3.6(a). Since the power of the input signal at the sample needs to be small ($P_{\mathrm{in}} \sim -85\,\mathrm{dBm}$), the source attenuation of the VNA is employed and additional attenuators (INMET 18AH) are installed in the input line (port 1). The attenuators are specified for room temperature and up to 18 GHz. However, tests showed that their performance does not decrease up to 20 GHz and at low temperatures. After being attenuated the signal then passes through the sample.

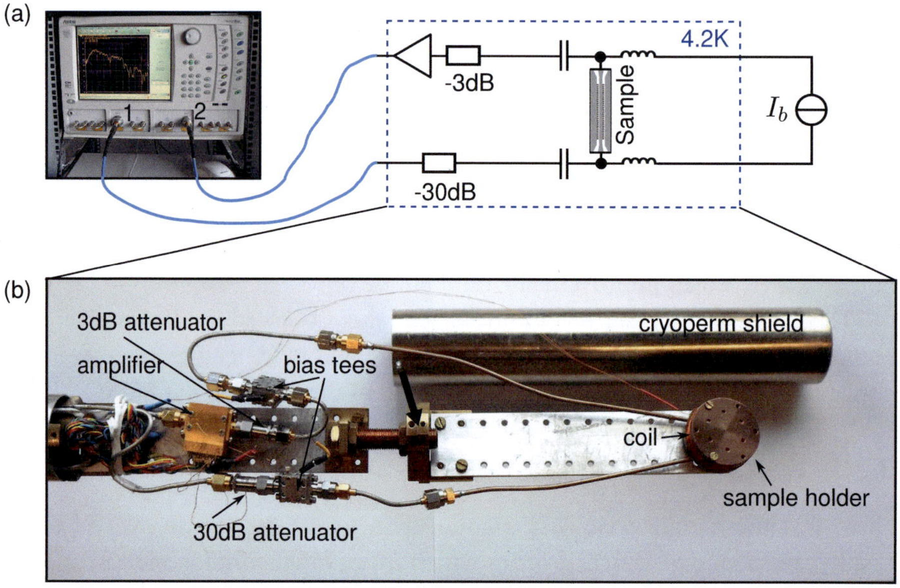

Fig. 3.6: (a) Sketch of the experimental setup together with a photograph of the VNA. (b) Photograph of the cold part of the electronic setup. The cryoperm shield is not yet installed, but the wide black arrow shows where it will be fixed and its horizontal position indicates how much of the setup will be covered.

xA Low Noise Factory amplifier (LNF-LNC6_20A) with a gain of 30 dB between 6-20 GHz is installed after the sample to amplify the output signal. Typically, there are reflections from the amplifier back into the sample. This parasitic effect is counteracted by installing either a small attenuator (3 dB) or a Pasternak circulator (PE8403 with a frequency range from 7 GHz to 12.4 GHz) between the sample and the amplifier. Although the mode of operation of the circulator allows a better performance than the attenuator, it emanates strong magnetic fields. Due to the high sensitivity of the SQUID meta-atom to magnetic field, it can therefore not be employed in the experiments. In order to shield the sample from other, weaker magnetic fields, e.g. the earth's magnetic field or fields due to weakly magnetic electric components, a cryoperm shield is installed around the sample. Cryoperm is a mu-metal, i.e. a metal with a high magnetic permeability ($\mu = 50000 - 140000$), that has good magnetic shielding properties at low temperatures. More details on the sensitivity and protection of the sample against magnetic field will be given in Sec. 3.3.

The blue rectangle in Fig. 3.6 indicates the part of the setup (including the sample) which is immersed in liquid helium at a temperature $T = 4.2\,\mathrm{K}$. Figure 3.6(b) shows a photograph of this part of the setup. The cylindric magnetic shield in the photograph is not yet attached but the wide black arrow indicates where it will be fixed.

Apart from the high frequency setup described to far, there is also a direct current (dc) component. A dc current source and Marki bias tees (BT-0025) with a range from 40 kHz to 25 GHz are used to superpose the microwave signal with a constant current. This current I_b in the central conductor creates a constant magnetic field which, as elucidated in Sec. 2.1.3, tunes the resonance frequency of the rf-SQUIDs.

3.2.3 Magnetic Field Bias

Apart from the magnetic field bias by means of the bias tees, there is also a coil around the sample holder (cf. Fig. 3.6). It creates a field perpendicular to the area of the chip, i.e., perpendicular to the area of the SQUID loop. Both magnetic fields can be used to tune the resonance frequency of the meta-atoms. The main difference between the two options is the symmetry of the field with respect to the gap as illustrated in Fig. 3.7. While the magnetic field in one gap is

$$B_{\mathrm{gap1}} = B_{\mathrm{stray1}} + B_{\mathrm{coil}} + |B_{\mathrm{cond}}|, \tag{3.3}$$

the field in the second gap is

$$B_{\mathrm{gap2}} = B_{\mathrm{stray2}} + B_{\mathrm{coil}} - |B_{\mathrm{cond}}|. \tag{3.4}$$

Here, only the y-component of the corresponding $\vec{B}$-field is considered. $\vec{B}_{\mathrm{stray}}$ takes into account any homogeneous fields which are not applied intentionally, e.g. the earth's magnetic field. Ideally, the sample is protected from such fields. However, depending on the quality of the magnetic shielding, stray fields may linger. In this case, the y-component of the stray magnetic field is not zero, and as a result there may be a small offset in magnetic field between the two gaps.

Then, the resonance curves of SQUIDs in different gaps do not lie on top of each other, which decreases the maximum possible signal. Depending on which means (bias tees or coil) is used for the magnetic field tuning of the SQUID resonance frequency, the other (coil or bias tees) has to be used to counteract the stray magnetic field. In Fig. 3.7 the example of magnetic field bias by the central conductor and the coil as opposition for the (homogeneous) stray magnetic field is shown.

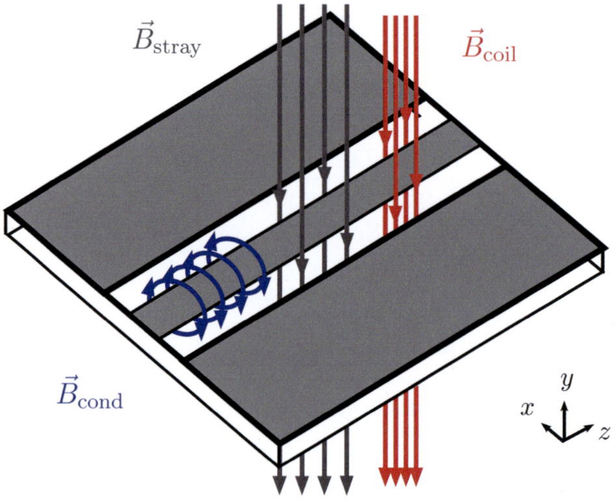

Fig. 3.7: Sketch of the magnetic field directions $\vec{B}_{\rm coil}$ (red) and $\vec{B}_{\rm cond}$ (blue) in the two gaps of the CPW. As an example, the stray magnetic field $\vec{B}_{\rm stray}$ (dark gray) is assumed to be homogeneous in the two gaps. The density and location of the arrows does not denote the respective field strength or distribution in space.

Which means is chosen for the magnetic field bias depends on the circumstances. Typically, the field produced by the dc current in the central conductor is homogeneous along the CPW on the chip, while the field of the coil decays slightly when moving away from the center. Hence, the magnetic field bias with the central conductor would be favorable to the one by the coil. However, if due to a inaccuracy in the fabrication process, the SQUIDs inside one gap are positioned closer to the central conductor than in the other, the coil is favorable, since its field across the whole chip is much more homogeneous than that of the central conductor.

3.2.4 Calibration Methods

Part of the content of this section is taken from Ref. [But+13a].

In Sec. 2.4.2, it was explained how the effective, relative permeability is retrieved from transmission data. For this procedure, it is assumed that the corresponding ports, i.e. the reference planes, of the scattering matrix are positioned directly in front of and behind the metamaterial. The transmission coefficient S_{21} measured by the VNA, however, comprises the transmission data through the full setup. Therefore, the measured data has to be calibrated properly to eliminate the effect of the setup.

In a real experiment, the best set of reference planes achievable for a full calibration of such a measurement is located at the microwave connectors closest to the sample at cryogenic temperatures. This, however, does not only require a more complex experimental setup [YA13], it is also insufficient for our retrieval method.

To solve this issue, measurement and simulation are combined to prepare the data for the retrieval algorithm. First of all, we divide the experimental setup into three parts, each of which can be described individually by a two-port scattering matrix:

- S^{in} describes the input part of the setup, from port 1 of the network analyzer to the beginning of the SQUID loaded section of the CPW.
- S^{stl} describes the SQUID loaded section of the CPW.
- S^{out} describes the output part of the setup from the end of the SQUID loaded section of the CPW to port 2 of the network analyzer.

We measure the total transmission from VNA port 1 to port 2 which can be written as

$$S_{21}^{\mathrm{tot}} = -\frac{S_{21}^{\mathrm{in}} S_{21}^{\mathrm{stl}} S_{21}^{\mathrm{out}}}{\left(S_{22}^{\mathrm{in}} S_{12}^{\mathrm{stl}} S_{21}^{\mathrm{stl}} - \left(S_{22}^{\mathrm{in}} S_{11}^{\mathrm{stl}} - 1\right) S_{22}^{\mathrm{stl}}\right) S_{11}^{\mathrm{out}} + S_{22}^{\mathrm{in}} S_{11}^{\mathrm{stl}} - 1}. \tag{3.5}$$

If we assume that the waveguides leading to the sample are well matched (i.e. no reflections at the waveguide connections to the sample: $S_{22}^{\mathrm{in}} = S_{11}^{\mathrm{out}} = 0$), the expression simplifies to

$$S_{21}^{\mathrm{tot}} = S_{21}^{\mathrm{in}} S_{21}^{\mathrm{stl}} S_{21}^{\mathrm{out}}. \tag{3.6}$$

Clearly, all of the above matrix elements are frequency dependent. S_{21}^{stl}, however, also depends on the dc magnetic field.

It can be decomposed into two factors by assuming $S_{21}^{\mathrm{stl}} = \tilde{S}_{21}^{\mathrm{stl}}(\omega) \cdot \alpha(\omega, \Phi_{\mathrm{e0}})$.

$\tilde{S}_{21}^{\mathrm{stl}}(\omega)$ is the transmission through the loaded waveguide section without the effect of the SQUID resonance. Consequently, this factor does not depend on magnetic field.

$\alpha(\omega, \Phi_{\mathrm{e0}})$ describes the change in the transmission through the loaded waveguide due to the SQUID resonance. Therefore, this factor is frequency and field dependent.

The total transmission then reads

$$S_{21}^{\mathrm{tot}}(\omega, \Phi_{\mathrm{e0}}) = S_{21}^{\mathrm{in}}(\omega)\, \tilde{S}_{21}^{\mathrm{stl}}(\omega)\, \alpha(\omega, \Phi_{\mathrm{e0}})\, S_{21}^{\mathrm{out}}(\omega)\,. \tag{3.7}$$

The goal of the calibration is to extract the term S_{21}^{stl} from the measured S_{21}^{tot} since the former is the quantity needed for the effective μ_r retrieval. If we restrict our investigation to a limited frequency range bounded by ω_{min} and ω_{max}, we can usually find a value of the flux $\Phi_{\mathrm{e0}} = \Phi_{\mathrm{cal}}$ for which $\alpha(\omega, \Phi_{\mathrm{cal}}) \approx 1$. Thus, by dividing all the measured data in the specified frequency range by the corresponding value at the calibration flux, we can extract α:

$$S_{21}^{\mathrm{tot,cal}}(\omega, \Phi_{\mathrm{e0}}) = \frac{S_{21}^{\mathrm{tot}}(\omega, \Phi_{\mathrm{e0}})}{S_{21}^{\mathrm{tot}}(\omega, \Phi_{\mathrm{cal}})} \approx \alpha(\omega, \Phi_{\mathrm{e0}}) \tag{3.8}$$

As a last step, we have to reconstruct S_{21}^{stl} from the calibrated data $S_{21}^{\mathrm{tot,cal}}$. Therefore, we simulate the transmission through the loaded waveguide without the SQUID resonance $\tilde{S}_{21}^{\mathrm{stl}}(\omega)$ using the finite element software "SONNET" and multiply it by the calibrated data:

$$S_{21}^{\mathrm{stl,reconstructed}}(\omega, \Phi_{\mathrm{e0}}) = S_{21}^{\mathrm{tot,cal}}(\omega, \Phi_{\mathrm{e0}}) \cdot \tilde{S}_{21}^{\mathrm{stl,simulated}}(\omega) \tag{3.9}$$

This result can then be used in the algorithm outlined in Sec. 2.4.2. It should be stressed, however, that this method is only valid if the effect of the SQUID resonance at the calibration flux is negligible in the frequency range of interest.

'thru' calibration

The easiest way to obtain the calibrated transmission data $S_{21}^{\mathrm{tot,cal}}$ is to use the built-in 'thru' calibration function of the VNA. This function effectively subtracts the reference data $S_{21}^{\mathrm{tot}}(\omega, \Phi_{\mathrm{cal}})$ in [dB] from the rest of the measurement in [dB]. For the reference data, an external flux value of $\Phi_{\text{cal}} = \pm 0.5\Phi_0$ is typically chosen. At this flux value, the resonance frequency of the SQUID meta-atoms is at its minimal value $\omega_{0,\text{min}}$. Sufficiently above the resonance frequency, the effect of the SQUIDs on the transmission through the CPW is small. Thus, using this calibration, $S_{21}^{\text{tot,cal}}$ is measured directly provided that the frequency range of interest is sufficiently (more than the bandwidth of the resonance) above $\omega_{0,\text{min}}$.

artificial calibration after the measurement

In the case that no calibration is used during the measurement, the subtraction of the data (in [dB]) measured at $S_{21}^{\text{tot}}(\omega, \Phi_{\text{cal}})$ $\Phi_{\text{cal}} = \pm 0.5\Phi_0$ can be also be done by hand after the measurement. This is often helpful to improve the clarity of the measurement result, since parasitic, not magnetic field dependent resonances are canceled. While it is possible to use the artificially calibrated data for the μ_r retrieval, this artificial calibration method will only be used in order to improve the data visibility.

As a last option, instead of using the data at a fixed flux value Φ_{cal}, the transmission data is averaged along the flux axis at each frequency point. This data is subsequently subtracted from the measured data. This option is used only in the next section in order to improve the clarity of data that has a statistical spread of resonance lines along the flux axis and it is not employed for quantitative analysis.

3.3 Protecting the Samples against Stray Magnetic Flux

Part of the content of this section is published in Ref. [But+13b].

3.3.1 Abrikosov Vortices

As already briefly outlined in Sec. 2.1.1, Abrikosov vortices are magnetic field penetrations in a superconducting material. They are tube-shaped with the axis oriented along the magnetic field lines. The superconducting order parameter is suppressed towards the axis of the tube and zero directly at the center. The magnetic flux trapped inside such a vortex always carries one flux quantum.

In a bulk type II superconductor, such as Nb, at temperature $T = 0$, they occur only once a critical magnetic field $H_{c1}(T = 0)$ is exceeded. However, this critical field decreases as the temperature approaches the critical temperature of the superconductor. Therefore, already small magnetic fields can cause Abrikosov vortices if the field is present while the superconductor is cooled from above to below the critical temperature. The vortices then remain trapped in the superconductor even as the temperature is further decreased and H_{c1} increased.

Additionally, the trapping is increasingly easier, the thinner the film becomes. On the other hand, decreasing the width (and not the length) of such a film, lessens the probability of trapping vortices [SFM04; PNS07]. In fact, Stan *et al.* [SFM04] showed that decreasing the width of a strip of Nb leads to an increased critical field $H_{c1}(T_c)$. If the sample is cooled from above to below T_c in a field $H < H_{c1}(T_c)$, no Abrikosov vortices occur. Hence, the smaller the width of the film, the less vortices are trapped. Additionally, vortices are trapped preferably at inhomogeneities in the superconductor [PNS07], for example at vias where two layers of superconductor connect.

3.3.2 The Effect of Stray Magnetic Fields

As explained in the previous chapter and the previous section, the samples used in this work are made from thin-film Nb structures. Especially the magnetic meta-atoms, i.e. the rf-SQUIDs, are extremely sensitive to any magnetic field. Already a field of about 1.2 μT creates a flux in the SQUID loop that equals one flux quantum.

An Abrikosov vortex creates a local distortion to the magnetic field which affects only SQUIDs in close vicinity to the vortex. This leads to an non-negligible offset in magnetic field between different SQUIDs. As a result, the resonance curves (cf. Fig. 2.3) of individual SQUIDs are shifted against each other.

In the case of a single meta-atom as discussed in Sec. 2.4.1 (cf. Fig. 2.11), the trapping of vortices is not a problem. It only shifts the resonance curve along the flux axis, without deforming it. In a symmetric arrangement, with one SQUID in each gap, flux inhomogeneities lead to two resonance curves that are shifted against each other. This effect can be counteracted using the antisymmetric magnetic flux bias of the bias tees to make the two curves overlap as explained in Sec. 3.2.3.
However, this is no longer possible for any larger number of SQUIDs if the stray flux is not homogeneous. Therefore, in order to observe a collective resonance curve of all SQUIDs in the sample, care has to be taken to avoid any stray magnetic field. First, if the stray magnetic field is large and inhomogeneous, it itself leads to a inhomogeneous flux bias of individual SQUIDs. But even if the field is small, it may cause Abrikosov vortices, which, in turn, destroy the field homogeneity. Such stray magnetic fields are either caused by magnetic components used in the experimental setup or external sources such as the earth's magnetic field. The first negative effect of the field itself may be counteracted by shielding the sample from any such fields as well as possible.

Additionally, the sample has to be designed in such a way as to discourage the trapping of Abrikosov vortices in order to suppress the second effect. This can be done by using normal metal instead of superconductors where possible and by decreasing the width of the superconducting structures in order to reduce their demagnetization factor, as mentioned above [SFM04].

sample description

In the following, the results of transmission measurements on two different samples M1 and M2 will be used to demonstrate the means necessary to obtain one collective resonance curve for (almost) all meta-atoms. Each sample contains 54 SQUIDs, 27 per gap. In sample M2, a fully superconducting waveguide made of Nb is employed. The SQUID meta-atom used in sample M2 is shown in Fig. 3.2. Its parameters are given in Tab. 3.3. An optical micrograph of the rf-SQUID of the second sample M1 is shown in Fig. 3.1(b). Its specific parameters are also given in Tab. 3.3. The general behavior of the two SQUIDs is the same. However, since their parameters differ, the SQUIDs in the two samples have a different minimal and maximal resonance frequency.

In order to suppress the occurrence of Abrikosov vortices in sample M1, two measures were taken. First, the ground planes of the CPW of sample M1 are made of normal metal (Pd) instead of Nb. The central conductor is still made from Nb, due to requirements of the fabrication process (voltage bias during anodization of the SQUIDs). Second, the SQUIDs used in M1 have a smaller width d of the superconducting leads and area of the via A (cf. Fig. 3.2 and 3.1). As explained above, this should considerably suppress the trapping of Abrikosov vortices in sample M1 due to the different design.

results

The field (bias current) and frequency dependent complex transmission S_{21} through the samples is measured using a setup similar to the one presented in Sec. 3.2.2. The main differences are that the complete cold part of the electronic setup is inside the cryoperm shield. Additionally, a circulator with a range from 7 GHz to 12.4 GHz is used instead of the 3 dB attenuator in front of the amplifier and flexible microwave cables connect to the sample holder. Contrary to its specifications the measurement shows that the performance of the circulator is not noticeably decreased up to 20 GHz.

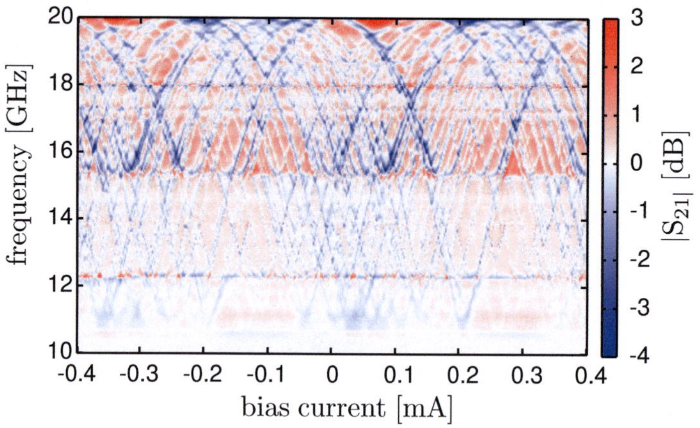

Fig. 3.8: Transmission magnitude through sample M2 in dependence on flux and frequency.

amplifier	bias tee	flexible microwave cable (10 cm)	attenuator	circulator (unshielded)
0.1	6	1	0.03	150

Table 3.5: Average magnetic fields created by various components in [μT]. All fields were measured at a distance of a few millimeters by an axial magnetic field probe inside a cylindric cryoperm shield.

When sample M1 is used to replace sample M2, an improved result is expected due to the different design. However, the resulting transmission again shows again a multitude of resonance lines (not shown here). Thus, further measures to improve the setup had to be taken.

All measurements except the one shown in Fig. 3.11 are calibrated using the average after measuring as explained in Sec. 3.2.4) only to improve clarity. The resulting transmission data for sample M2 in the initial setup is presented color coded in Fig. 3.8. Decreased transmission is shown in blue. The picture shows many lines spread randomly over the full current range which is proportional to the magnetic flux. Each line corresponds to the magnetic flux dependent resonance curve of one or a small number of rf-SQUIDs. The shift between these lines reflects the inhomogeneity of magnetic flux across the sample. The horizontal lines at approximately 12.2 GHz, 15.5 GHz and 18 GHz are parasitic sample holder resonances that couple to the SQUIDs and distort the resonance lines.

All electronic components were examined with an axial magnetic probe (Teslameter FM302 by Projekt Elektronik) to determine if and how strongly they are magnetic.

The magnetic field probe is positioned inside and parallel to the axis of a vertically oriented cylindrical cryoperm shield as shown in Fig. 3.6. The probe rests approximately 5 cm above the bottom of the shield, which has a total length of 20 cm. The offset magnetic field in the empty shield amounts to about 15 nT. Each component is then lowered with a nonmagnetic string into the shield until it rests a few millimeter below the lower end of the probe. Subsequently, the magnetic field due to the electronic component is measured. The resulting value depends on the orientation of the component with respect to the probe and not all orientations are tested. The measured values are listed in Tab. 3.5.

The circulator is by far the component with the strongest magnetic field, which cannot be sufficiently screened by the cryoperm shield anymore. It is therefore replaced by an attenuator as mentioned already in the previous section. Irreplaceable are the amplifier and the bias tees which, unfortunately, are also magnetic. In order to protect the sample from their fields, they a placed outside the cryoperm shield.

How these measures improve the behavior of the SQUID metamaterial sample M1 is shown in Fig. 3.9: The spread of the lines is strongly reduced.

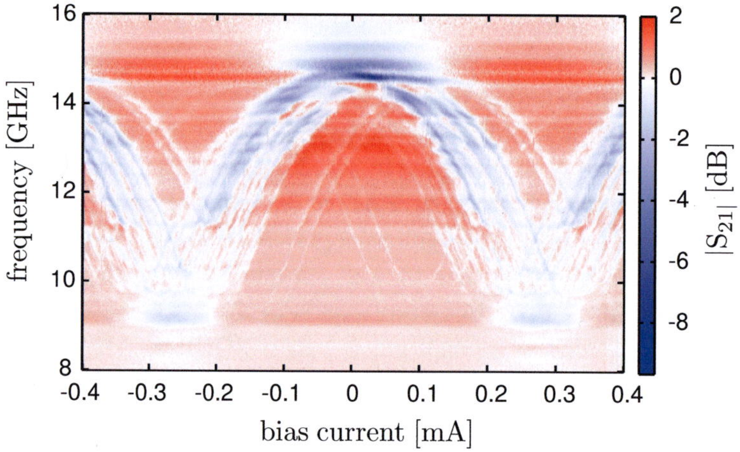

Fig. 3.9: Flux and frequency dependent transmission magnitude through sample M1 with improved magnetic environment: The circulator is replaced by a 3 dB attenuator and amplifier and bias tees are installed outside the cryoperm shield.

Yet, the overlap of the lines is still not good enough. There are two reasons for that. First, the sample is positioned in the upper half of the cryoperm shield, i.e. close to the magnetic electronics. Second, the flexible microwave cables used to connect to the connectors in the sample holder are also magnetic (cf. Tab. 3.5). Since the cables cannot be removed or installed outside the cryoperm shield, either care has to be taken that they pass by the sample without affecting it or they have to be replaced by non-magnetic, semi-rigid cables, e.g. pure copper or brass. Interestingly, many microwave cables, that seemingly consist only of copper contain ferromagnetic material in the inner core of the central conductor.

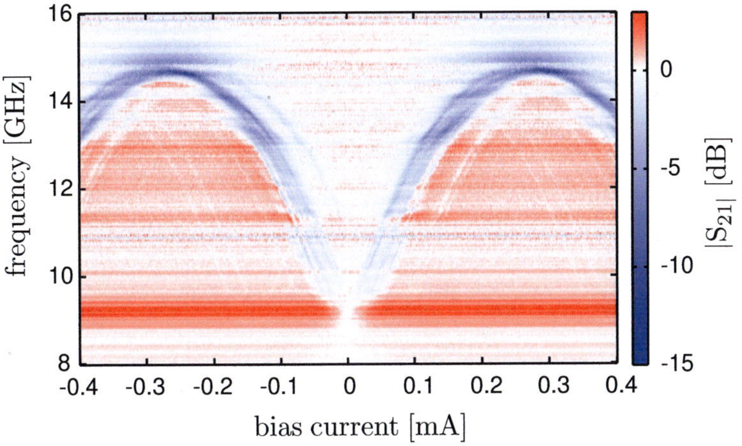

Fig. 3.10: Transmission magnitude through sample M1. Compare to Fig. 3.9, the sample is moved deeper into the shield (lower third) and the microwave cable passes at the side of the sample.

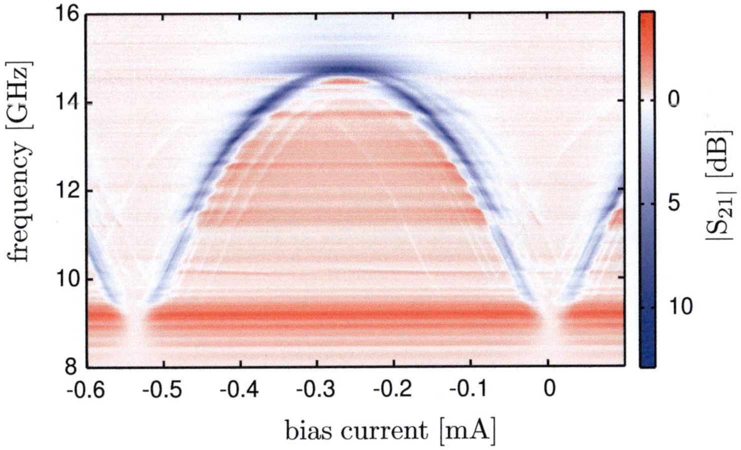

Fig. 3.11: Transmission magnitude through sample M1 in the improved sample holder. The microwave cable does not pass the sample anymore. Note that the calibration for this measurement was done in situ, using the through calibration function of the VNA at an external flux $\Phi_{e0} = 0.5\Phi_0$, i.e. at a bias current of $-0.53\,\text{mA}$.

The comparison of Figs. 3.9, 3.10 and 3.11 shows how the magnetic environment is considerably improved by careful arrangement of the cables. In Fig. 3.9, one cable passes on top of the sample holder with a distance of approximately 10 mm to the sample. In Fig. 3.10, this cable is moved to pass along the side of the sample and the sample is moved deeper into the cylindrical cryoperm shield. It now rests in the lower third of the shield. In Fig. 3.11, a different sample holder, namely sample holder S1, at the same position deep inside the cryoperm shield is used. Although the shape of the inner volume of the sample holder stays the same, it is rotated by 90° around its axis. This allows for the microwave cables to be led away from the sample without passing it. Additionally, it leads to an orientation of the waveguide on the chip that is perpendicular to the axis of the magnetic shield.

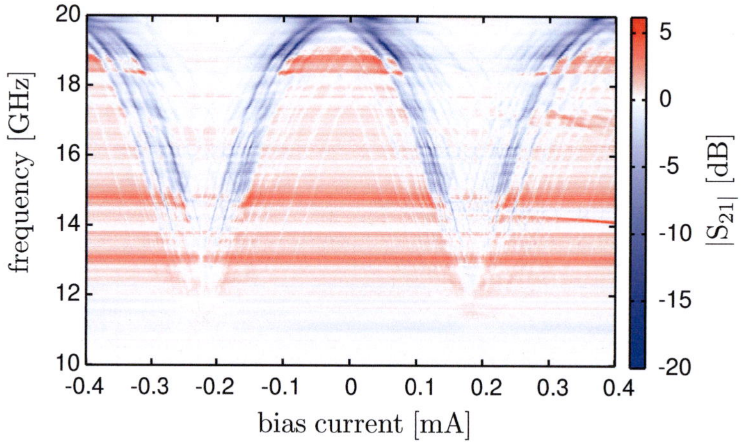

Fig. 3.12: Transmission magnitude through sample M2 with the same setup used in the measurement shown in Fig. 3.11.

The strength of the magnetic field that is residual in the cryoperm shield is not constant along the axis of the shield while almost constant in the radial direction. Therefore, this new orientation of the waveguide decreases the spread in magnetic field along one gap. On the other hand, the field bias of the two gaps may differ slight. This effect, however, can be counteracted by the dc current in the central conductor (cf. Fig. 3.7). Finally, with this improved setup, one main resonance curve is clearly visible. As expected, the same result (shown in the next chapter) is obtained, when the microwave cables are replaced by non-magnetic ones instead of just carefully arranging them.

From Figs. 3.9 to 3.11, it becomes clear, that protecting the sample from magnetic fields is crucial. Even the magnetic fields of microwave cables affect the quality of the result. It should be emphasized, that the result in Fig. 3.11 was obtained with sample M1.

We now change back to sample design M2. When it is installed in exactly the same setup as the one used for the measurement on sample M1 in Fig. 3.11, the quality of the result is degraded again (see Fig. 3.12). The main difference between the two samples M2 and M1 is their affinity to trap Abrikosov vortices. Thus, the vortices are the reason for the degradation of the performance of sample M2 under otherwise identical measurement conditions as for sample M1.

conclusion

Before a tunable magnetic metamaterials as described in Sec. 2.4.2 could be investigated, one main challenge had to be overcome. Magnetic components in the setup and trapped Abrikosov vortices give rise to an inhomogeneous magnetic environment.

In order to avoid this, an optimized positioning of sample, electronic components and even coaxial cables as well as proper magnetic shielding is crucial. Additionally, the trapping of Abrikosov vortices has to be prevented. Therefore, superconducting planes, e.g. large ground planes, have to be avoided, if possible. Moreover, wide structures in the rf-SQUID have to be reduced in one dimension to discourage the trapping of vortices, also the area of contact between two superconductors needs to be reduced as much as possible. By applying all these measures, a collective, tunable resonance curve of almost all 54 rf-SQUIDs was achieved.

4 Experimental Results

In this chapter, experimental results obtained with a purely magnetic and a purely electric one-dimensional metamaterial are presented. For the magnetic medium, flux and frequency dependent transmission data is shown and discussed. From the transmission coefficient, the effective magnetic permeability of the magnetic medium is determined. Transmission data is also presented for the electric metamaterial. Additionally, finite element simulations are used to further understand its behavior.

4.1 A One-Dimensional Magnetic Metamaterial

The one-dimensional magnetic metamaterial consists of 54 rf-SQUIDs as magnetically interacting elements. In the following, the experimental results obtained with sample M1 are presented and discussed.

4.1.1 Transmission Measurements at 4.2 K

initial measurement

Sample M1 is installed in the experimental setup and cooled to liquid helium temperature, $T = 4.2\,\mathrm{K}$. Subsequently, a first measurement of the transmission coefficient over a wide range of frequency and magnetic flux is produced. Therefore, a fixed current is applied to the coil, while the VNA is used to sweep n_{avg} times the given frequency range, which is divided into n_{freq} points. The resulting frequency dependent transmission coefficient S_{21} is obtained by averaging over the n_{avg} sweeps. Whenever the frequency of the input signal is the same as that of a resonance within the setup, a dip in the transmission magnitude and a change in the transmission phase is observed. The frequency sweeps including the averaging are then repeated for n_{flux} coil current values.

The result for the transmission magnitude $|S_{21}|$ in [dB] with respect to the input signal ($P_{\mathrm{input}} = -15\,\mathrm{dBm}$) of such a measurement is shown in Fig. 4.1. The frequency range from $2 - 20\,\mathrm{GHz}$ is divided into $n_{\mathrm{freq}} = 300$ points and the current axis contains $n_{\mathrm{flux}} = 800$ values within a current range between -80 and $+80\,\mu\mathrm{A}$. Each trace corresponding to a fixed current value is averaged twice.

The result is then used to determine the range of tunability, i.e. the frequency range of interest, which lies between 8 and 16 GHz for sample M1. Additionally, the periodicity in coil current is used to determine the current corresponding to one flux quantum. For this sample, one Φ_0 corresponds to a magnetic field of $1.2\,\mu\mathrm{T}$ and equates to a coil current of $53.2\,\mu\mathrm{A}$. This value is used to normalize the current or, respective, flux axis with respect to Φ_0.

The measurement in Fig. 4.1 is not calibrated. The power in [dBm] at the sample is

$$P_{\mathrm{sample}} = P_{\mathrm{input}} + P_{\mathrm{Satt}} + P_{\mathrm{cold\ att}} + P_{\mathrm{coax}} \approx -95\,\mathrm{dBm}. \qquad (4.1)$$

$P_{\mathrm{Satt}} = -40\,\mathrm{dB}$ is the source attenuation of the VNA, $P_{\mathrm{cold\ att}} = -30\,\mathrm{dB}$ the cold attenuation, and $P_{\mathrm{coax}} \approx -10\,\mathrm{dB}$ is the attenuation in the coaxial cables.

Due to this strong attenuation ($\approx -80\,\mathrm{dB}$ in total), the transmission magnitude is comparatively low although the cold amplifier ($\approx +30\,\mathrm{dB}$) is used. Resonances, in this setup, are visible as dips (white or blue color) in the transmission magnitude. Any resonance line with a flux independent resonance frequency is a parasitic resonance of the setup. These lines are calibrated out in later measurements. However, any resonance that produces a magnetic field perpendicular to the SQUID area couples to the SQUID resonance curves, thus affecting their shape. This effect cannot be calibrated out.

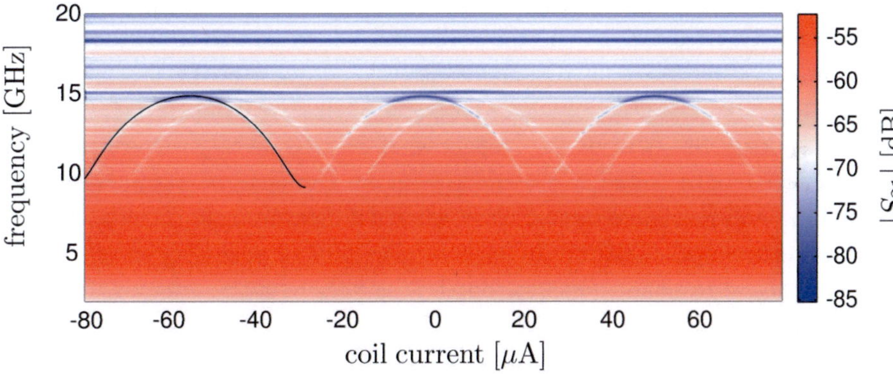

Fig. 4.1: Uncalibrated transmission magnitude through sample M1 containing 27 rf-SQUIDs in each of the two gaps. The transmission is presented in color code in dependence of frequency and bias current through the external coil. The black line indicates the flux dependent resonance frequency calculated from the sample parameters.

In Fig. 4.1, two resonance lines are visible that change their frequency with changing coil current, i.e. magnetic flux. These are the resonance curves of the 1D metamaterial on sample M1, which consists of 2×27 rf-SQUIDs. The offset between the two lines is due to an offset in magnetic flux between the SQUIDs in both gaps. As pointed out in Sec. 3.2.3, the reason for this offset is the residual magnetic field in the cryoperm cylinder that changes along the vertical axis of the shield. A closer look at both SQUID resonance curves reveals that the decrease in transmission is not the same for both lines although each gap contains the same number of meta-atoms. This is for two reasons. First, due to slight uncertainties in the fabrication process, the distance between central conductor and the chain of SQUIDs in one gap is less than in the other. Hence the coupling to the microwave is weaker for one line than for the other.

Second, the chip containing sample M1 is not symmetric, it contains a second transmission line. Simulations showed, that the SQUIDs in one gap couple, i.e. radiate, more strongly to the sample holder resonance at 15 GHz, which further decreases their signal.

The black line in Fig. 4.1 following the most pronounced curve between $-80\,\mu$A and $-30\,\mu$A shows the calculated flux dependent resonance curve. It is determined by using the parameters of sample M1 as given in Tab. 3.1 with Eq. (2.17). The calculated flux dependent position of the resonance frequency agrees very well with the measured data.

As explained in Sec. 3.2.3, the two resonances are superposed by applying a small magnetic field via the central conductor. All other, hardly visible, resonance lines correspond to the resonance curves of single SQUIDs that are offset in magnetic flux most probably due to trapped Abrikosov vortices. The vortices distort the magnetic field of the respective rf-SQUID locally. Unfortunately, this effect cannot be completely avoided with the available measurement setup. These lines will be visible more clearly when the measurement is calibrated.

After the appropriate current in the central conductor necessary to superpose the two relevant resonance lines is found, it suffices to concentrate on the relevant frequency range and only one period of Φ_0.

high resolution transmission measurement

Figure 4.2 shows the section ranging from $-0.6 < \Phi_{e0}/\Phi_0 < 0.6$ in the flux axis and from 8 GHz to 16 GHz in the frequency axis with higher resolution in both axes ($n_{\text{freq}} = 801$ and $n_{\text{flux}} = 800$). The measurement is calibrated using the 'thru' calibration of the VNA (cf. Sec. 3.2.4) at $\Phi_{e0}/\Phi_0 = -0.5$. Since a dip is subtracted around 9 GHz, a transmission larger than 0 dB is observed in this frequency range at flux values where the resonance frequency of the SQUIDs is higher.

The transmission on resonance is reduced by up to 11 dB as shown in color scale in Fig. 4.2. Due to the calibration, the aforementioned resonance lines of single SQUIDs are now clearly visible as white or light blue lines. These SQUIDs do not contribute to the collective resonance. Nonetheless, the remaining SQUIDs, approximately 50, show one common resonance curve. Small deformations of the resonance line due to the coupling to parasitic resonances are most pronounced for example at $\nu \approx 14.5\,$GHz and $\nu \approx 13.6\,$GHz.

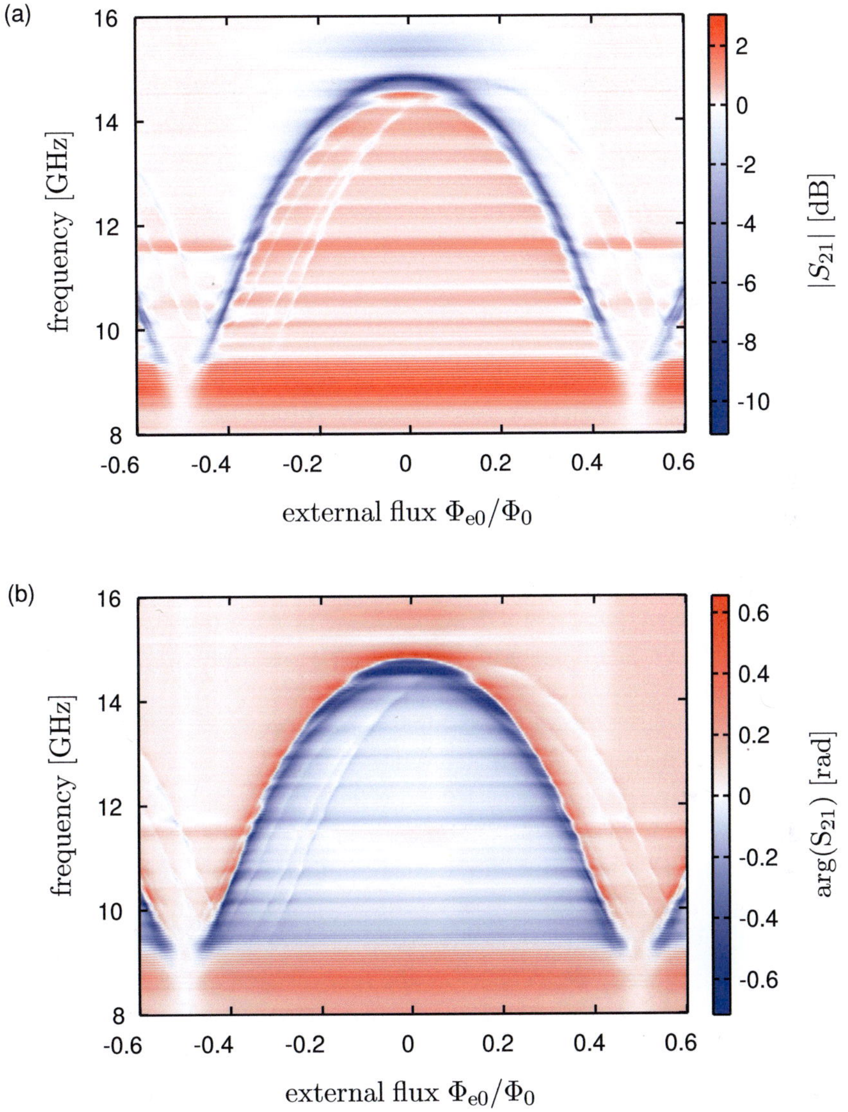

Fig. 4.2: Calibrated complex transmission. (a) Frequency and flux dependent transmission magnitude and (b) phase through sample M1 containing 54 rf-SQUIDS.

Figure 4.2(b) shows the corresponding transmission phase also in color scale. It is defined as the phase difference between input and output signal. At resonance, the phase changes quickly from negative to positive values. As for the magnitude, around 9 GHz, the "subtracted" resonance due to calibration is visible.

A closer look at the 2D plot of the phase far away from resonance reveals that the phase drifts slightly during the measurement from left to right. The phase signal of the VNA is not completely stable. Since a measurement takes several hours, this small instability is observable as a small drift in the data. Additionally, the phase drifts because the helium level in the dewar decreases slowly during the measurement which affects the temperature gradient along the coaxial cables. This, in turn, changes the phase along the cables which is mapped onto the metamaterial due to calibration. Both effects lead to a very small increase of the phase from negative to positive flux. This will be visible again later, when the magnetic permeability of this system is calculated.

single traces

A more detailed look at transmission magnitude and phase is possible if only the frequency dependence at a fixed flux value and in a narrower frequency range around the resonance frequency is considered. The traces of $|S_{21}|$ and $\arg(S_{21})$ at flux values $\Phi_{e0}/\Phi_0 = 0$ (red), -0.1 (blue), -0.17 (pink) and -0.21 (green) are shown in Fig. 4.3(a) and (b), respectively.

The transmission magnitude shows the expected dip with a decrease of about -10 dB. All but the pink curve ($\Phi_{e0}/\Phi_0 = -0.17$) show side minima or maxima. These are the effect of parasitic resonances that couple to the SQUIDs. Nonetheless, the shift of the position of the resonance with changing magnetic flux is indisputable. The corresponding behavior is observed in Fig. 4.3(b), which shows the frequency dependence of the transmission phase for the same flux values. The expected steep phase change at resonance is clearly visible, as is, again, the effect of parasitic resonances.

Although the shape of the resonances is far from perfect, the transmission magnitude can be used to fit a Lorentzian of the form

$$A(\omega) = \frac{A_0\gamma^2}{(\omega - \omega_0)^2 + \gamma^2} + 1 \tag{4.2}$$

to the data. Therefore, the measured linear data in $[V]$ is inverted and squared, in order to show a positive peak and to obtain an energy scale. A_0 is the amplitude of the Lorentz curve, ω_0 the position of the resonance and 2γ the full width at half maximum. The $+1$ is added to account for the offset of $+1$ of the data from zero. The result of the fit is shown in Fig. 4.3(a) with thin black lines on top of the corresponding resonance curves. From the Lorentzian, a result for the different quality factors $Q = \omega_0/(2\gamma)$ is extracted and given in Tab. 4.1. The errors represent the 95% confidence bounds.

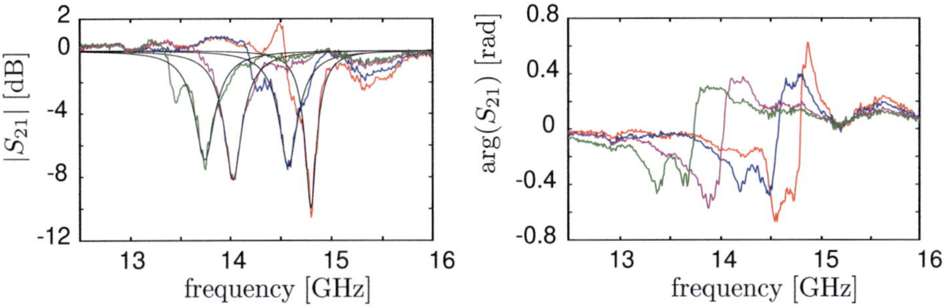

Fig. 4.3: Frequency dependence of transmission (a) magnitude and (b) phase at normalized external flux values $\Phi_{e0}/\Phi_0 = 0$ (red), -0.1 (blue), -0.17 (pink) and -0.21 (green). The thin black lines in (a) correspond to the Lorentzian fit result.

Except for the resonance at zero flux, these quality factors are comparable to the quality factor obtained by fitting Eq. (4.2) to the resonance curve of one of the stray single SQUID lines pointed out previously. In this case, the line that has a resonance frequency of 13.56 GHz at a flux of $\Phi_{e0}/\Phi_0 \approx 0.38$ is used.

The extracted valuc for onc singlc rf-SQUID is $Q = 94 \pm 13$. The error is so large because the signal to noise ratio for the single SQUID resonance is rather poor.

Since the quality factor of the collective resonance is comparable to the quality factor of the single SQUID resonance, we can conclude, that there is no synchronization between the meta-atoms. Due to the small coupling between nearest neighbor SQUIDs which is about 1/20 of the coupling to the transmission line, this is expected.

Φ_{e0}/Φ_0	0 (red)	-0.1 (blue)	-0.17 (pink)	-0.21 (green)
Q	188 ± 8	96 ± 4	101 ± 3	82 ± 4

Table 4.1: Quality factors and errors obtained from a Lorentzian fit to the transmission data shown in Fig. 4.3(a). The errors represent the 95 % confidence bounds.

4.1.2 Magnetic Permeability at 4.2 K

With the knowledge of the complex transmission $S_{21} = |S_{21}| \exp(i \cdot \arg(\mathrm{S}_{21}))$, we are now able to retrieve the relative magnetic permeability of the SQUID metamaterial. As start frequency for the retrieval process $\nu_{\min} = 10\,\mathrm{GHz}$ is chosen, in order to ensure that the effect of the 'thru' calibration at $9\,\mathrm{GHz}$ is as small as possible. Using the procedure explained in Sec. 2.4.2, the complex, relative magnetic permeability of the sample is calculated.

As mentioned in Sec. 2.4.2, finding the correct root relies strongly on and is sensitive to the initial parameters. A stable approach was found by using the solution of the previous frequency point as the initial value for the root finding algorithm of the next higher frequency point. To find the initial values at the lowest frequency point $\nu_{\min}$, it is assumed that the SQUIDs have no effect at that frequency and corresponding start values, namely $\mu_r = 1$, $S_{11} = S_{22} = 0$ and $S_{12} = S_{21}$ are chosen. The resulting real part of the relative magnetic permeability is shown in Fig. 4.4(a), the imaginary part, which corresponds to losses, in Fig. 4.4(b). Both are depicted between $-0.6 < \Phi_{\mathrm{e}0}/\Phi_0 < 0.6$.

In Sec. 2.4.2, two possibilities as to how to test the reliability of the result were explained. Here, only the phase delay method is used. Therefore, the phase delay $\tau_{\delta,\mathrm{calc}}$, which is calculated from the retrieved data (Fig. 4.5(a)), is compared to the measured phase delay $\tau_{\delta,\mathrm{meas}}$ (Fig. 4.5(b)). The difference of the two values is given in Fig. 4.5(a). As required for the correct solution, the difference between the two values is close to zero.

Now, that the validity of the calculated relative magnetic permeability is ensured, a closer look at the results is justified. First of all, Fig. 4.4(a) and (b) clearly show how the magnetic permeability is tunable with magnetic flux as expected from the transmission data. The color scale of Fig. 4.4(a) shows that the real part of μ_r varies between 0 and 2.6. However, these values are not reached for every frequency. This becomes even more clear when looking at single traces. Figure 4.6(a) and (b) show real and imaginary part of μ_r at flux values $\Phi_{\mathrm{e}0}/\Phi_0 = 0$ (red), -0.1 (blue), -0.17 (pink) and -0.21 (green). The colors and the flux values are the sames as for the single traces shown in Fig. 4.3.

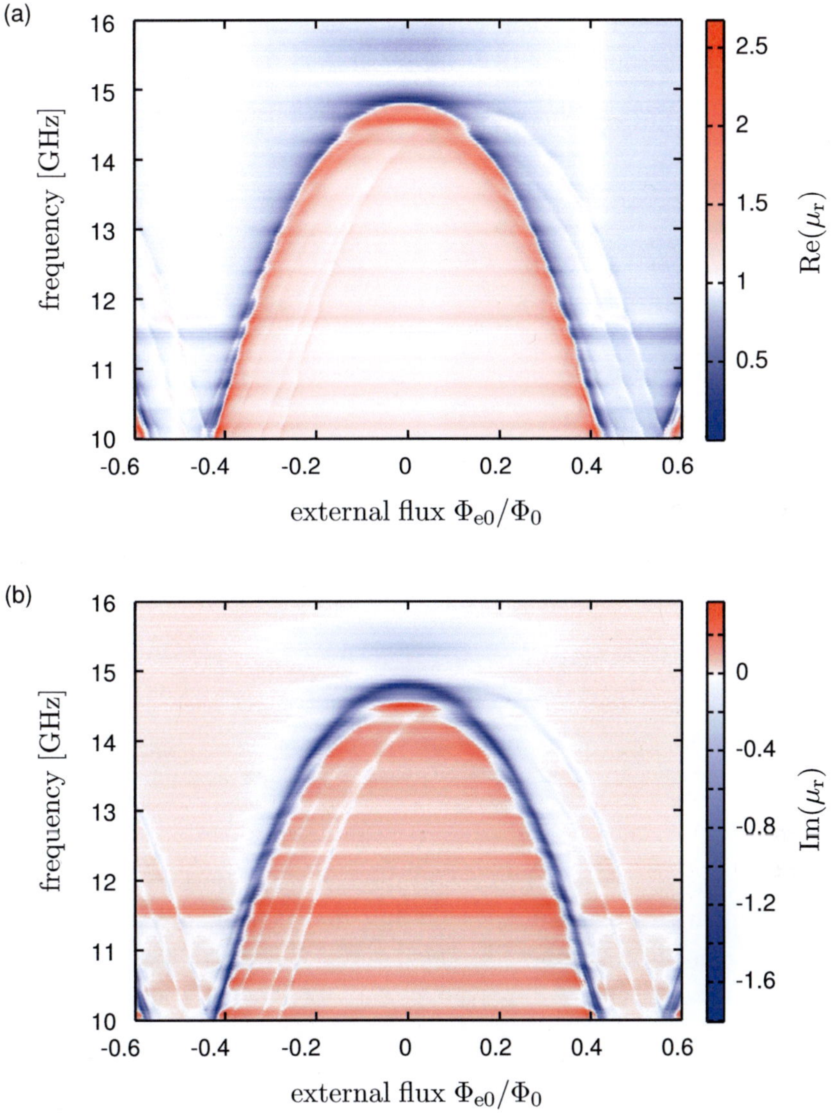

Fig. 4.4: (a) Real part of the relative magnetic permeability calculated from the transmission data shown in Fig. 4.2. (b) Imaginary part of μ_r. Note that $\mu_r = \mathrm{e}(\mu_\mathrm{r}) + \mathrm{iIm}(\mu_\mathrm{r})$, instead of the commonly used $\mu_r = \mu_r' - i\mu_r''$.

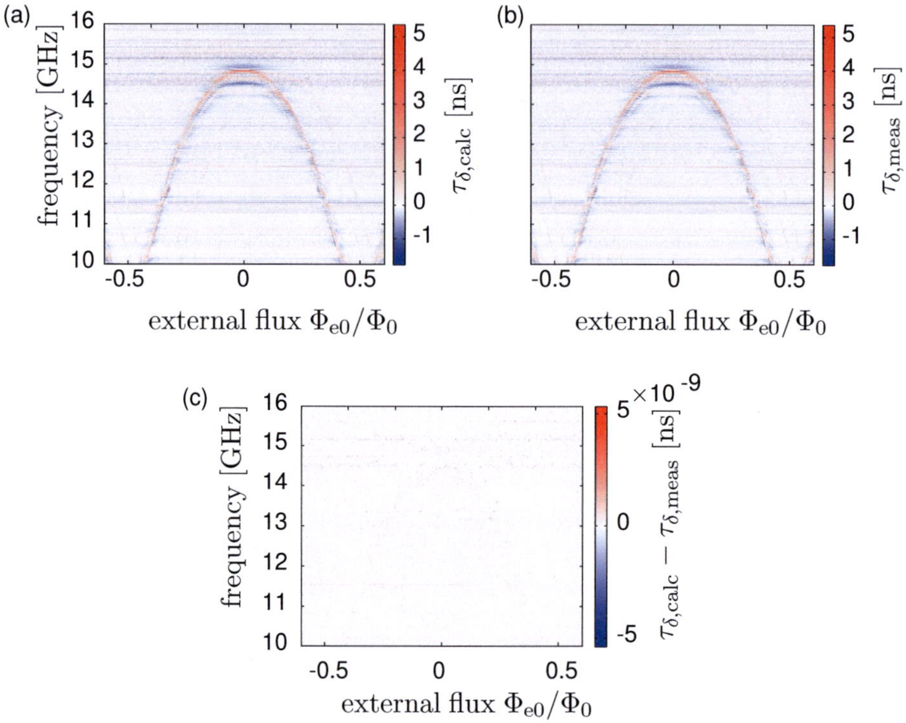

Fig. 4.5: Comparison of the (a) calculated and (b) measured phase delay τ_δ. (c) Difference between the calculated and measured phase delay. Note the different scaling of the color scale.

None of the depicted curves in Fig. 4.6(a) reaches zero. Nonetheless, as predicted, by changing the external magnetic flux, the magnetic permeability can be tuned. The tunability is further illustrated in Fig. 4.7(a) and (b).

Figure 4.7(a) shows how, at a frequency of 12.96 GHz (red) and 14.01 GHz (blue), any value between 0.3 and 1.6 can be chosen for the real part of μ_r simply by changing the external flux Φ_{e0} applied to the metamaterial.

Instead of changing the magnetic permeability at one fixed frequency, one can also use the flux tunability to have the same value of $\mathrm{Re}(\mu_\mathrm{r})$ at different frequencies. This is demonstrated in Fig. 4.7(b). The blue dots indicate the flux values at which the real part of the relative magnetic permeability is 0.6 for the corresponding frequency. Usually, this value is reached twice for one flux value. Once directly at or very close to the resonance, where losses are large, and a second time, when $\mathrm{Re}(\mu_\mathrm{r})$ increases to unity again (cf. Fig. 4.6(a)). Since losses are less at that second frequency value, this higher frequency is used in the plot. Figure 4.7(b) also shows that the value $\mathrm{Re}(\mu_r) = 0.6$ is not reached for every frequency. The jumps at $\Phi_{e0}/\Phi_0 \approx \pm 0.06$ and ± 0.15 are an artifact of parasitic resonances as can be seen from Fig. 4.2(a). To summarize the experimental results: Although a metamaterial was created that clearly shows a tunable relative magnetic permeability, the real part of this permeability never becomes negative.

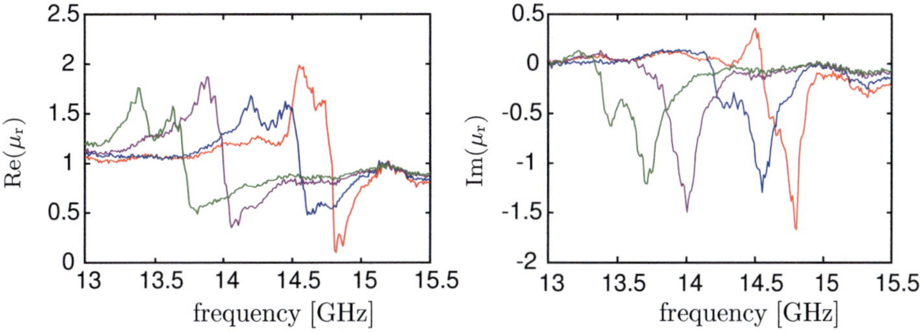

Fig. 4.6: Frequency dependence of the (a) real and (b) imaginary part of the magnetic permeability at flux values $\Phi_{e0}/\Phi_0 = 0$ (red), -0.1 (blue), -0.17 (pink) and -0.21 (green) corresponding to the single traces of transmission data shown in Fig. 4.3.

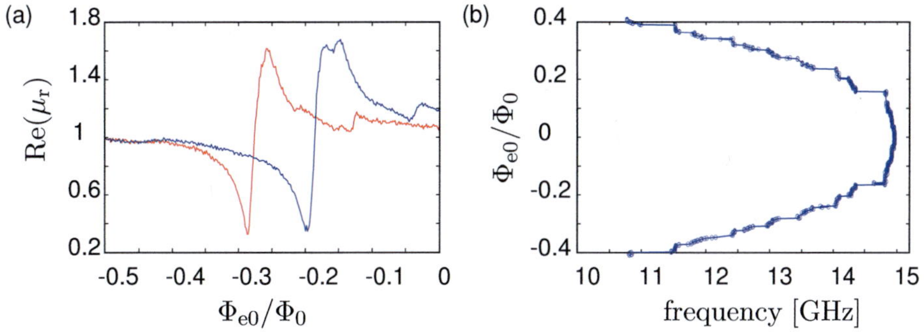

Fig. 4.7: (a) Flux dependent real part of the magnetic permeability at a frequency of 12.96 GHz (red) and 14.01 GHz (blue). (b) The blue dots indicate the flux values that need to be chosen in order to obtain $\mathrm{Re}(\mu_r) = 0.6$ at the corresponding frequency. The blue line is a guide to the eye.

In Fig. 2.8 in Sec. 2.3.2, it was shown that the quality factor of the resonance is connected to the magnetic susceptibility of the resonator. Therefore, a closer look at the quality factor shall be taken. For all n_{flux} resonance curves in the measurement, Eq. (4.2) is fitted to data in order to obtain the respective quality factor. The result is shown in Fig. 4.8(a) in blue.

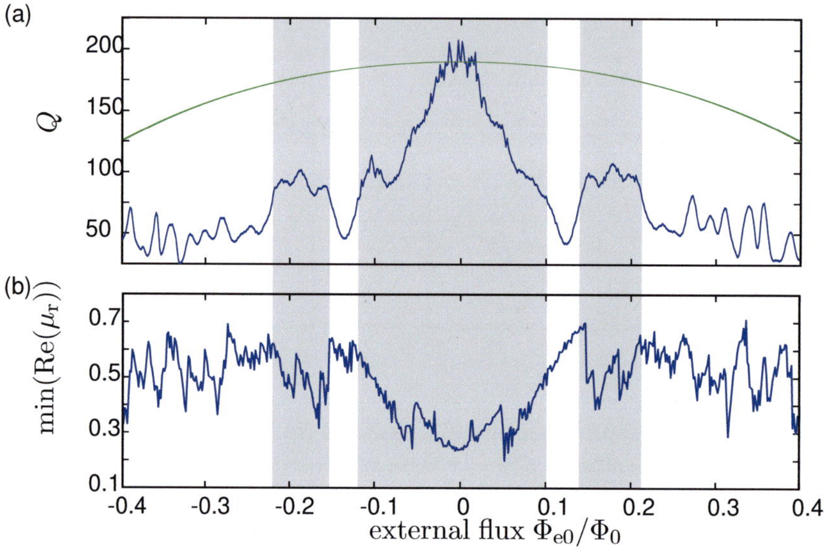

Fig. 4.8: (a) Calculated quality factor Q_{calc} (green) of the single SQUID and experimental results Q_{exp} obtained from ~ 54 SQUIDs (blue). (b) Minimal value of $\text{Re}(\mu_{\text{r}})$ at different flux values. The gray areas indicate flux values that have a $Q_{\text{exp}} > 75$ and correspondingly smaller minimal values of $\text{Re}(\mu_{\text{r}})$.

For comparison, the theoretical flux dependence of the quality factor Q_{calc} is also shown in green. The quality factor of a parallel RLC circuit is given by $Q_{\text{calc}} = 1/(\omega_0 RC)$. In terms of the rf-SQUID this can be written as $Q_{\text{calc}} = \omega_c \omega_0 / \omega_p^2$. The frequencies describing the damping ω_c and junction plasma frequency ω_p are defined in Sec. 2.1.2, ω_0 is the flux dependent resonance frequency. Since the resonance frequency ω_0 changes with external flux while ω_c and ω_p are constants, the quality factor decreases as Φ_{e0} deviates from zero.

This dependence on magnetic flux is in general also observed in the measured data, Q_{exp}. On the other hand, the flux dependence of the experimental result for the quality factor is clearly non-monotonic. The reason therefore is that the experiment does not reflect a constant γ, i.e. ω_p^2/ω_c, due to a non-ideal electromagnetic environment. Additionally, the comparison of Q_{exp} and Q_{calc} shows, that the experimental values stay mostly below the theoretical value which is the quality factor of the single SQUID. First, again the parasitic resonances in the sample holder have to be named as one origin. Second, it may be that the superposition of the SQUIDs with respect to the flux axis is not perfect. This explains the conclusion that no synchronization is observed as already stated earlier.

Let us return to the connection between quality factor and magnetic susceptibility. A decreased quality factor leads to a decreased deviation from zero of the susceptibility in either positive or negative direction. In turn, for a material consisting of many such resonators, the deviation from unity of the relative magnetic permeability decreases. This connection between minimal value of $\mathrm{Re}(\mu_{\mathrm{r}})$ and the quality factor Q of the resonance is also found in our measurements. The quality factor is compared to the corresponding minimal value of $\mathrm{Re}(\mu_{\mathrm{r}})$ (cf. Fig. 4.8(b)). Since the result for the quality factor is a fit result, this curve is smoother than $\min(\mathrm{Re}(\mu_{\mathrm{r}}))$ which is extracted directly from the measured data.

The gray areas indicate flux values where a quality factor above 75 is observed. In these areas, the real part of the relative magnetic permeability typically assumes low values. On the other hand, there are also deviations from this observation. Especially at flux values that deviate by more than $0.25\Phi_0$ from zero flux, the correlation is no longer unambiguous. This is because not only the quality of the resonance, i.e. the transmission magnitude, affects the magnetic permeability, but also the variation in phase. The abrupt changes in $\min(\mathrm{Re}(\mu_{\mathrm{r}}))$ at $\Phi_{\mathrm{e}0}/\Phi_0 \approx \pm 0.06$ and ± 0.15 are due to the coupling of the meta-atoms to parasitic resonances.

Nonetheless, from Fig. 4.8 one can infer that increasing the quality factor of the metamaterial's collective resonance will lead to smaller and maybe even negative values of the magnetic permeability.

4.1.3 Transmission and Magnetic Permeability at 2.1 K

The critical temperature of bulk Nb is $T_c = 9.2\,\mathrm{K}$. For thin films, the critical temperature is in general decreased but for film thicknesses above $\sim 200\,\mathrm{nm}$, as in our case, this decrease is negligible [Gub+05]. Thus, at 4.2 K, the sample is at a temperature less than half its critical temperature, where the energy gap is already very close to that at zero temperature.

However, the density of quasiparticles, i.e. single electron like excitations, is not zero and changes with temperature. Together with the Cooper pairs, they can be described by a two-fluid model [Tin04], which is reflected in a complex conductivity. The quasiparticles yield the real part σ_1 (resistive contribution), while the Cooper pairs are responsible for the imaginary part σ_2 (inductive contribution). Both fluids are driven by the oscillating magnetic field. The quality factor of the corresponding resonance is proportional to the ratio σ_2/σ_1 (in the dirty superconductor limit, which is applicable in our case) [Bar09]. Since the density of quasiparticles decreases with decreasing temperature by several orders of magnitude between $0.5 > T/T_c > 0.1$, so does σ_1 [Bar09]. As a result (the density of superconducting electrons stays approximately constant), the ratio σ_2/σ_1 and hence the quality factor increases with decreasing temperature.

In order to achieve a better quality factor of the metamaterial's resonance, the same setup as before is used but the sample is cooled to 2.1 K by pumping on the liquid helium. Since the dissipation in all components is decreased, the sensitivity of the measurement is increased. While this is in principle positive, the measurement is also more sensitive to parasitic effects. For example, the small period oscillations in the frequency axis around the resonance frequency due to reflections are more pronounced. The reflections are a result of the mismatch of the SQUID loaded transmission line at resonance and 50 Ω. In order to minimize this effect, 3dB attenuators are installed directly in front of and behind the sample. Additionally, the phase stability of the whole setup (not only the VNA) is disturbed by two factors: the mechanical vibrations caused by the pumping and the changing temperature gradient due to a helium level that decreases faster than in the 4.2 K measurements. As we will see, this affects the result for the transmission phase. Since the time for this measurement is limited, only the flux range between $-0.3 < \Phi_{e0}/\Phi_0 < 0.3$ and the corresponding relevant frequency range is measured.

transmission coefficient

The measurment was also calibrated using the 'thru' calibration of the network analyzer at a flux of $\Phi_{e0}/\Phi_0 = -0.5$. The power at the sample is reduced to $P_{\text{sample}} \approx -100\,\text{dBm}$. The resulting frequency and flux dependent transmission magnitude and phase are depicted in Fig. 4.9(a) and (b), respectively.

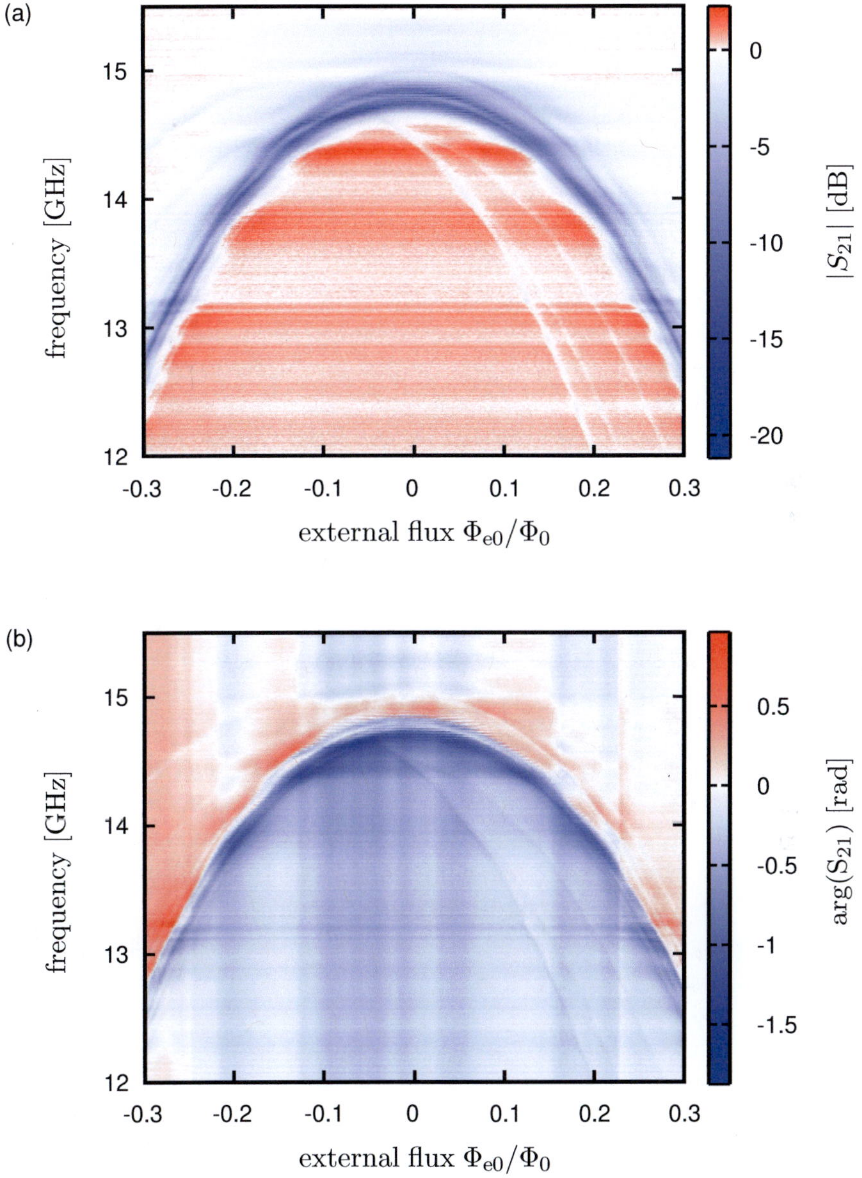

Fig. 4.9: (a) Transmission magnitude of measured on sample M1 at a temperature of 2.1K̇. Due to the enhanced quality factor, the decrease in transmission reaches down to -21 dB. (b) Transmission phase.

Again, the magnetic flux tunable resonance frequency is clearly observable. A closer look shows that there seem to be two resonance lines above 14 GHz and that the coupling of the lower curve to the parasitic resonance at 14 and 14.5 GHz, i.e. its deformation, is stronger. The reason for the latter was already mentioned in the previous subsection. Since the SQUIDs in one gap are positioned closer to the central line than the SQUIDs in the other gap, they couple differently to parasitic resonances. Additionally, the SQUIDs in different gaps could have slightly different parameters, e.g. critical currents, which would lead to different maximal resonance frequencies.

Due to the decreased sensitivity at higher temperature, this effect is not resolvable in the measurement at 4.2 K. At that temperature, the transmission magnitude never drops below -11 dB while at 2.1 K, it reaches down to -21 dB.

The transmission phase in Fig. 4.9(b) also shows the typical resonance behavior and the splitting of the resonances. Additionally, the phase is not constant during the measurement. Since the phase is very sensitive to smallest changes in the experimental setup, this is not surprising considering, that the data was recorded while pumping on the helium dewar. Due to the calibration, any phase change along the full experimental setup is mapped onto the short length of the loaded transmission line. This yields a measured phase drift that is not connected to physical properties and behavior of the metamaterial. Hence, the effect of this phase instability has to be counteracted. From Sec. 2.3.2, we remember, that the susceptibility of the single resonant meta-atom is close to zero at frequencies sufficiently below the resonance frequency. The same is valid for the transmission phase. Therefore, the phase is normalized by subtracting the average of the first 20 measurement points (12.0 GHz - 12.1 GHz) from each trace. This effectively defines to phase at 12 GHz to be zero.

As was already done in the previous subsection, the quality factor of the individual resonance curves is obtained by fitting a Lorentz distribution to the measured data. Four examples at different flux values $\Phi_{e0}/\Phi_0 = 0$ (red), -0.18 (blue), -0.24 (pink) and -0.26 (green) are depicted in Fig. 4.10(a). The fit results are again shown as thin black lines on top of the corresponding measured curve. The results for the quality factor are given in Tab. 4.2.

Figure 4.10(b) shows the corresponding transmission phase. The phase instability is corrected by subtracting the average of the first 20 points as explained above.

Due to the splitting of the resonance at higher frequencies and due to the oscillations at low transmission magnitude, the result of the fit at $\Phi_{e0}/\Phi_0 = 0$

does no longer reproduce the corresponding data (red). This is also reflected in the low quality factor and comparatively large error (cf.Tab. 4.2). The best quality factor in Tab. 4.2 is found at $\Phi_{e0}/\Phi_0 = -0.26$. Around this flux value, the best (i.e. negative) results for μ_r are expected.

Φ_{e0}/Φ_0	0 (red)	-0.18 (blue)	-0.24 (pink)	-0.26 (green)
Q	93 ± 13	142 ± 13	140 ± 10	206 ± 9

Table 4.2: Quality factors and errors obtained from a Lorentzian fit to the transmission data shown in Fig. 4.3(a). The errors reflect again the 95 % confidence bounds of the fit result.

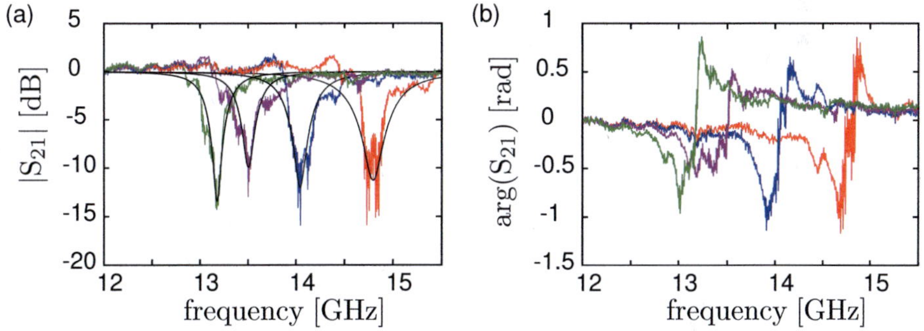

Fig. 4.10: (a) Transmission magnitude at flux values $\Phi_{e0}/\Phi_0 = 0$ (red), -0.18 (blue), -0.24 (pink) and -0.26 (green) and fit results. (b) Corresponding phase with corrected phase drift.

magnetic permeability

The relative magnetic permeability μ_r is obtained by the procedure described in Sec. 2.4.2 and the validity of the result is again confirmed by the phase delay comparison.

A brief look at the color scale of Fig. 4.11(a) confirms that the real part of the magnetic permeability indeed becomes negative and even reaches -1. As guide to the eye, the black lines in Fig. 4.11(a) represent the contour of $\mathrm{Re}(\mu_r) = 0$. Hence, negative permeability does not prevail for all resonance frequencies.

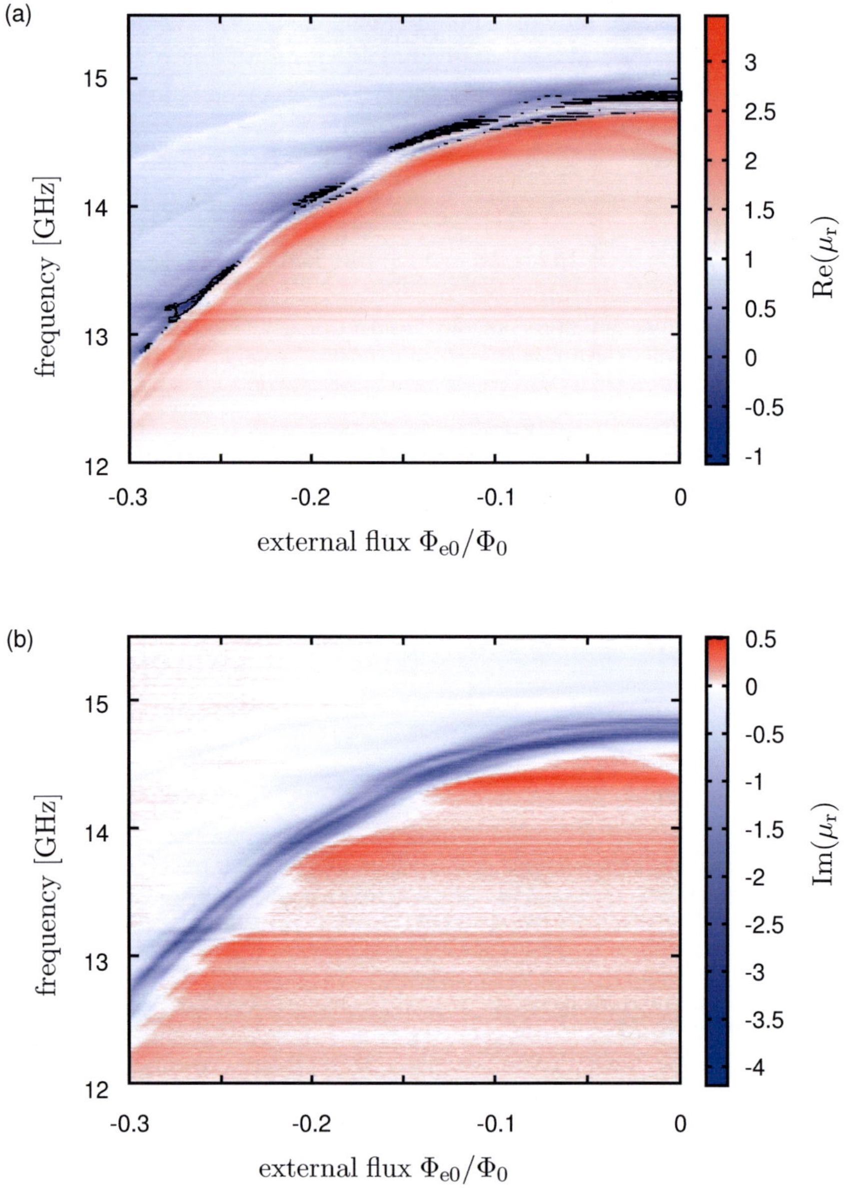

Fig. 4.11: Flux and frequency dependent (a) real and (b) imaginary part of the relative magnetic permeability calculated from the data measured at 2.1 K and presented in Fig. 4.9. The black lines in (a) indicate the $\mathrm{Re}(\mu_r) = 0$ contour.

The results for real and imaginary part of the relative permeability are shown in Fig. 4.11(a) and (b), respectively. Only the flux interval $-0.3 < \Phi_{e0}/\Phi_0 < 0$ is shown to improve the visibility of details. The result for μ_r was obtained using the corrected phase data. Therefore, both real and imaginary part of μ_r are homogeneous along the flux axis for frequencies sufficiently far away from the flux dependent resonance frequency.

Figure 4.11(b) shows the imaginary part of the relative magnetic permeability. As expected, the decreased transmission magnitude on resonance comes at the cost of increased losses in the material. The imaginary part of μ_r becomes as small as -4.2.

Since $\mathrm{Im}(\mu_r)$ is not relevant for the further discussion, only single traces of the real part are shown in Fig. 4.12 for flux values $\Phi_{e0}/\Phi_0 = 0$ (red), -0.18 (blue), -0.24 (pink) and -0.26 (green).

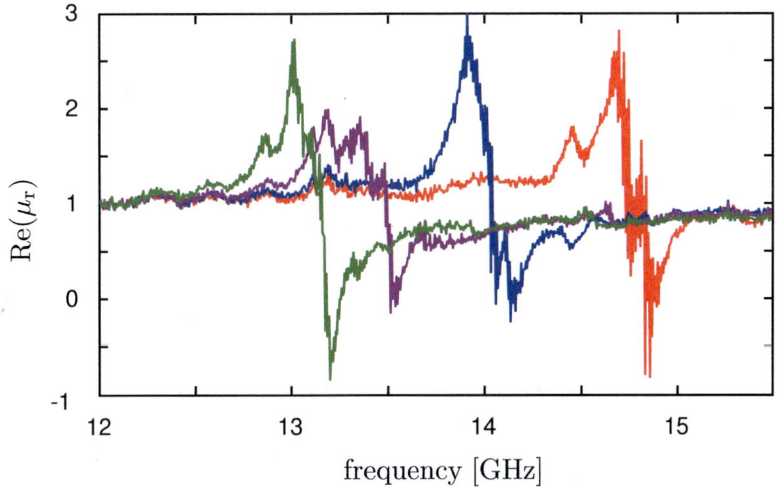

Fig. 4.12: Frequency dependent single traces of the real part of the relative magnetic permeability at flux values $\Phi_{e0}/\Phi_0 = 0$ (red), -0.18 (blue), -0.24 (pink) and -0.26 (green). The curves correspond to the single traces of transmission magnitude and phase shown in Fig. 4.10.

Comparing the green and the red curve illustrates a peculiarity of the measurement. While the green curve clearly inhibits a negative magnetic permeability in a small but finite frequency range, the red curve reaches negative values only at two singular frequency points. This is also reflected in the $\mathrm{Re}(\mu_r) = 0$ contour in Fig. 4.11(a). At frequencies above 13.5 GHz, the contour never encloses an area but rather indicates singular, fixed frequency values that seem to be periodic in the frequency axis. In fact, the periodic oscillations are a result of internal reflections due to the impedance mismatch once the SQUIDs are in resonance with the incoming signal. Including the aforementioned 3 dB attenuation before and after the sample reduced the oscillation amplitude but did not eliminate it completely.

In Fig. 4.10(a), the effect of the oscillations is strongest for the curve at zero flux. Considering the steepness of the decrease in transmission magnitude at the onset of the resonance suggests that the real decrease in transmission is obscured by these standing wave patterns (and the splitting of the resonance peak). This means that $\mathrm{Re}(\mu_\mathrm{r})$ would probably become much lower at $\Phi_{\mathrm{e}0} = 0$ without these oscillations.

4.1.4 Discussion & Conclusion

The experimental results presented in this section demonstrated the tunable resonance frequency of a transmission line metamaterial consisting of 54 rf-SQUIDs as magnetic meta-atoms. Using the retrieval method introduced in Sec. 2.4.2, it was shown that the tunable resonance frequency leads to a tunable relative magnetic permeability.

The measurements at 4.2 K showed a relative magnetic permeability, with a real part ranging between $0 < \mathrm{Re}(\mu_\mathrm{r}) < 2.6$. Although the tunability of the magnetic permeability was demonstrated for all frequencies between 10 GHz and 14.5 GHz, the minimal and maximal values of $\mathrm{Re}(\mu_\mathrm{r})$ could not be reached for every frequency. By comparing the quality factor of the individual resonance curves to the corresponding minimal values of the real part of μ_r, a correlation between large quality factor and low minimal value of $\mathrm{Re}(\mu_\mathrm{r})$ was found.

Since a decreased density of quasiparticles leads to an enhanced quality factor, the transmission coefficient was measured at a decreased temperature of $T = 2.1$ K. As expected, the range of $\mathrm{Re}(\mu_\mathrm{r})$ increased to vary between $-1.0 < \mathrm{Re}(\mu_\mathrm{r}) < 3.6$. Again, the minimal and maximal values were not observable at every frequency.

Instead, the frequency range between 13.18 GHz and 13.28 GHz, where the magnetic permeability was less than zero in a finite frequency range was small compared to the full range of tunability ($\sim$ 5 GHz) of the metamaterial. However, this frequency range was on the order of or even larger than the full range of frequency tunability reached for other tunable magnetic meta-atoms. For superconducting SRRs a frequency range of 50 MHz was achieved at a frequency of 10.77 GHz [RA06]. SRRs loaded with varactor diodes reached a larger range of about 500 MHz at 2.5 GHz [SMK06], the range of light tunable varactor loaded SRRs is considerably smaller, $\sim$ 20 MHz at 2.3 GHz. Additionally, for neither of these meta-atoms and corresponding materials the magnetic permeability was determined.

The reason, why a negative magnetic permeability like at $\Phi_{e0}/\Phi_0 = -0.26$ was not reached in the full range of tunability, was not a physical but a technical one. By further decreasing the parameter spread of the SQUIDs and by improving the sample holder, i.e. the electromagnetic environment for the sample, a negative permeability should be observable in the full 5 GHz range of tunability.

4.2 A One-Dimensional Electric Metamaterial

In this section, the experimental results measured on sample E1 are presented and discussed. Sample E1 contains 27 electric meta-atoms that couple capacitively to the coplanar waveguide. As for the magnetic meta-atom, transmission magnitude and phase are measured using the setup introduced in Sec. 3.2.2, however with less attenuation. The power at the sample is $P_{\text{sample}} \approx -88\,\text{dBm}$.

4.2.1 Measurement results and Comparison to Calculations

As elucidated in Sec. 2.4.3, the capacitors that couple to ground and central conductor of the CPW are connected by a superconducting rod that is interrupted by a small dc-SQUID and a single Josephson junction in series. Due to the small loop area, the dc-SQUID is considered to behave like a single junction but with a magnetic field tunable critical current. Thus, it serves as tunable inductor. Its coupling to the magnetic component of the microwave signal is negligible.

The impedance of the single rod structure, that spans both gaps, was found and used in Sec. 2.4.3 (in combination with the ABCD matrix of the unloaded line) to calculate the transmission coefficient S_{21}. The frequency and flux dependent result of the transmission magnitude in [dB] is depicted in Fig. 4.13. This figure is the same as Fig. 2.15 in Sec. 2.4.3. Since it will be helpful to compare calculated and measured results, it is shown here again.

The measurement procedure is the same as for the magnetic metamaterial, except that the central conductor is used for the magnetic field bias instead of the coil. From the periodicity in the horizontal axis, the current necessary for one Φ_0 is found to be $12\,\text{mA}$ and used to normalize the flux axis. The loop of the dc-SQUID is only 1/17 of the area of the loop of rf-SQUID which is used as magnetic meta-atom, therefore a larger magnetic field is necessary for one flux quantum.

In Fig. 4.13, the calculated and measured transmission magnitude is shown in the flux interval $-0.52 < \Phi_{e0}/\Phi_0 < 0.52$ for two different frequency ranges. Figure 4.13(a) shows the calculated result between $1 - 20\,\text{GHz}$, in Fig. 4.13(b) the calibrated experimental results are shown for a smaller frequency range between $2 - 12\,\text{GHz}$. The measurement was calibrated using the 'thru' calibration, while the sample (and amplifier) were at room temperature. At that temperature, the superconducting central conductor and the junctions behave purely resistive. As a result, the rods are no longer resonant in the depicted frequency range and

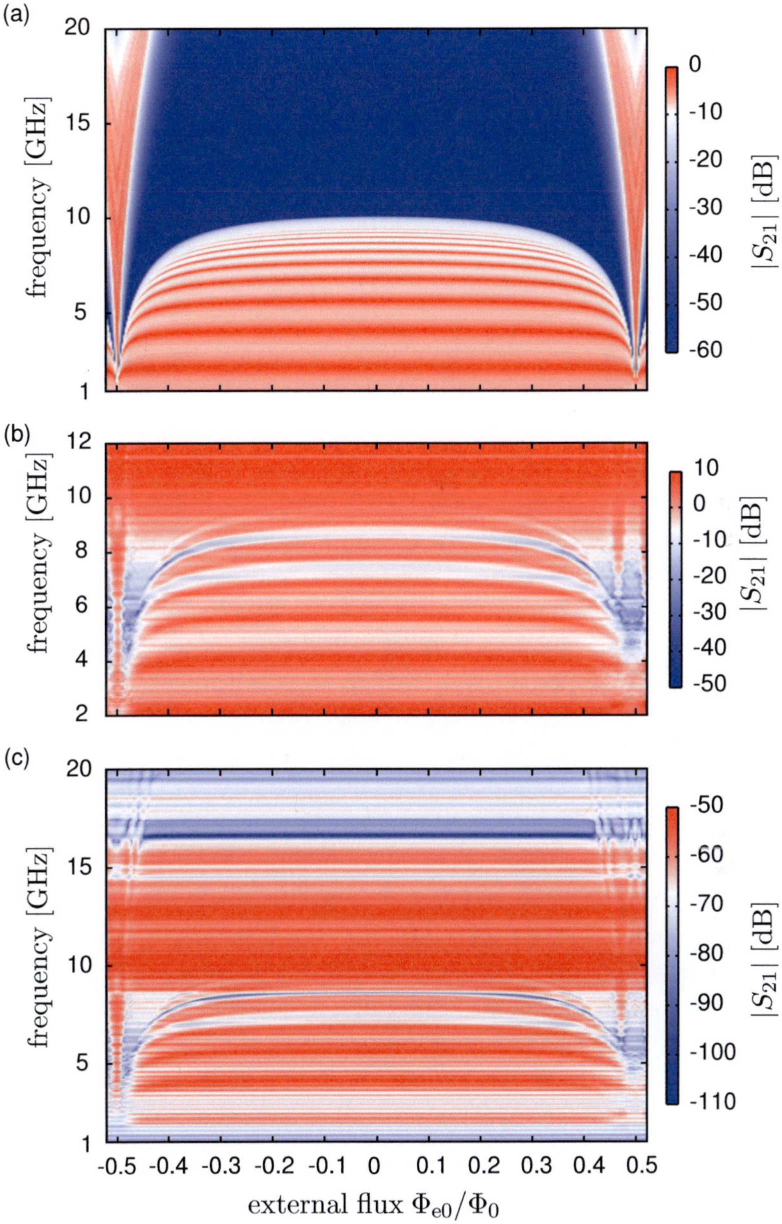

Fig. 4.13: Comparison of (a) calculated, (b) measured and calibrated and (c) measured and uncalibrated flux and frequency dependent transmission magnitude through sample E1 consisting of 27 rods.

their effect on the transmission is purely dissipative. However, at room temperature parasitic resonances and reflections in the setup are still present although less pronounced. Therefore, their effect is partially calibrated out. The maximum of the transmission magnitude in the calibrated measurement is above zero since the transmission improves due to decreased dissipation once the sample is at 4.2 K. Figure 4.13(c), finally, shows the uncalibrated measurement result again in the frequency range $1 - 20\,\mathrm{GHz}$. We will concentrate on the features that are common to experimental and calculated results first and on the discrepancies later.

Comparing Fig. 4.13(a) and Figs 4.13(b),(c), shows that both, measurement and calculation, show similar features in the lower frequency range between 1 GHz and 9 GHz. For the comparison with the calculation, the calibrated data in Fig. 4.13(b) is used. Both, calculation and measurement, show large period oscillations along the frequency axis. They are due to finite size aspects of the metamaterial as discussed in Sec. 3.2.2. In addition, the position of the peaks (and dips) is shifted to lower frequencies once the flux deviates from zero. Also common to both plots are the features close to $\Phi_{e0}/\Phi_0 = \pm 0.5$ that span the full frequency range. They correspond to the onset of the second pass band. As the external flux in the dc-SQUID loop approaches $\pm\Phi_0/2$, its inductance approaches infinity. Therefore, the resonance frequency of the rod and the cutoff frequency of all bands decreases to zero. From these common features we can conclude, that our description of the frequency and flux dependence of the metamaterial is in principle correct.

Returning to the uncalibrated data shown in Fig. 4.13(c), this leaves the question, why there is a strong discrepancy between calculation and measurement in the frequency range from 9 GHz to 20 GHz. The discrepancy is visible in more detail when comparing the measured (red) and calculated (blue) frequency dependent transmission magnitude at zero flux as depicted in Fig. 4.14.

Starting with the measured curve (red), small period oscillations are visible on top of the large period oscillations. While the latter are due to finite size effects of the material, the small period oscillations are most probably the effect of reflections at the mismatched electrically loaded waveguide. The origin of these reflections and why they disappear above 10 GHz is unclear. The period of 240 MHz and related half wavelength of 90 cm does not correspond to any typical length in the setup directly. Further investigations would be necessary to identify the origin of these reflections. However, they do not affect the behavior of the electric metamaterial and can be calibrated out.

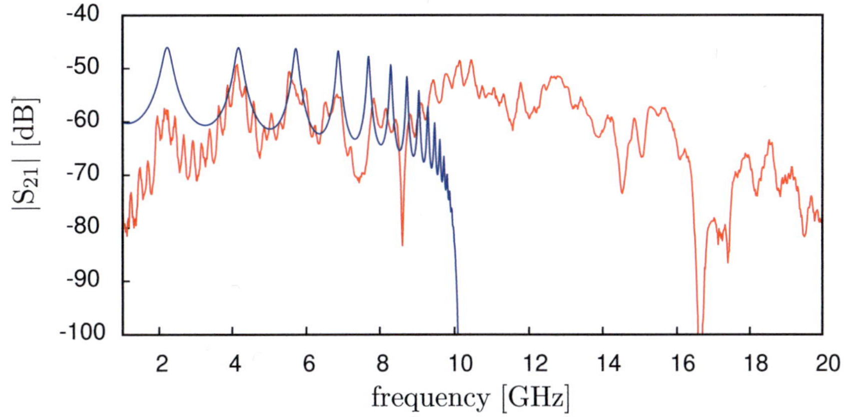

Fig. 4.14: Comparison of the calculated (blue) and measured (red) transmission magnitude at $\Phi_{e0}/\Phi_0 = 0$. The calculated curve is shifted by $-46\,\mathrm{dB}$ to facilitate the comparison.

For better comparison, the calculated curve is shifted by $-46\,\mathrm{dB}$, which is approximately consistent with the value of the attenuation minus the amplification. Note, that the decrease in measured transmission magnitude at low frequencies is due to the performance of the amplifier which is specified only down to 6 GHz. At frequencies below 6 GHz, the amplification decreases. Nonetheless, the finite size effect in the passband, which allows good transmission only at specific frequencies, is reproduced by the measurement and in good agreement with the calculation up to 9 GHz. However, the expected strong decrease in transmission above $\approx 10\,\mathrm{GHz}$ due to the onset of the stop band is not observed. On the contrary, the measured transmission goes slightly up instead of strongly down. In order to understand this behavior, the three-dimensional setup of the sample holder together with PCB and chip was simulated using the frequency domain solver of CST Microwave Studio [BTW13].

4.2.2 CST Simulations and Comparison to Measurement

The three dimensional design of the sample holder S2, the PCB and the Si chip (cf. Sec. 3.2.1) are loaded as 3D model into CST Microwave Studio. The coplanar waveguide on the chip is included as a 2D structure. In order to simplify the problem, the copper of the sample holder and the metallic layer on the chip are modeled as perfect conductor. The two copper layers of the PCB, however, are considered to be real copper. The silicon of the chip and the ceramic layer in the PCB are modeled lossless. Figure 4.15(a) and (b) show the top and the bottom part of the model, respectively. The parts of the sample holder are in beige (light brown), the PCB is shown in yellow, the chip again in beige. The blue stripes around the chip are the wire bonds used to connect the sample to the PCB. The two red planes that are oriented perpendicular to the waveguide on the PCB symbolize the input and output waveguide ports.

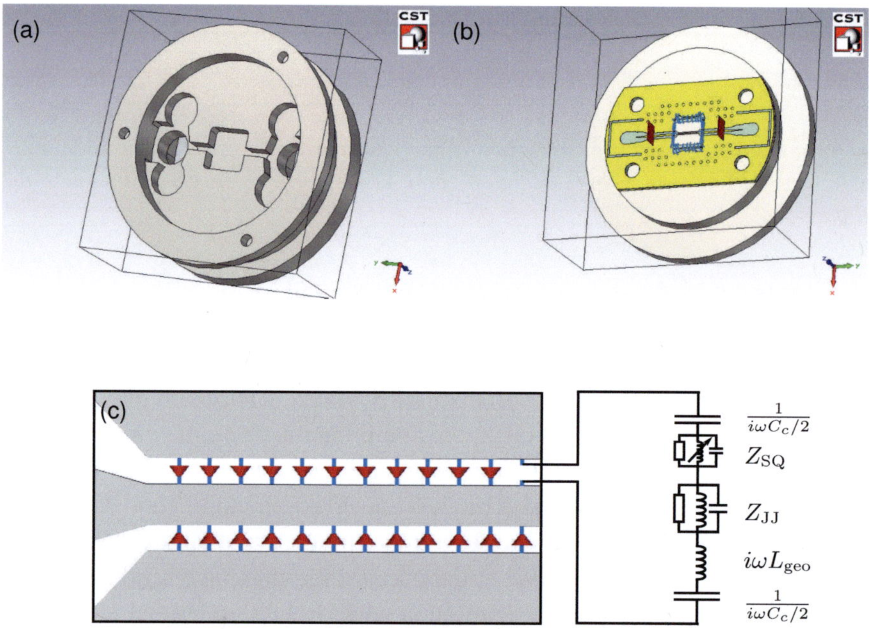

Fig. 4.15: 3D model of the sample holder (a) top part and (b) bottom part including PCB, chip and bond wires. (c) Zoom to the waveguide loaded with differential ports including a schematic of the SPICE model included between the two ends of the differential port.

CST Microwave Studio simulates 3D electromagnetic high frequency fields. Therefore in frequency domain, port 1 is driven by a QTEM mode of the respective frequency. By using a finite element method, the spatial distribution of the fields is simulated. The result is then used to calculate the magnitude of the signal transmitted from port 1 to port 2. CST can also be employed to find the spatial mode structure at defined frequencies, for example at sample holder resonances.

In order to simulate the frequency dependent response of the rod loaded transmission line, 2×27 additional differential ports were included into the CPW on the chip. Any port, apart from the input and output port (red planes in Fig. 4.15(b)), always has two ends. For a standard port, in contrast to differential, the potential at one end of the port is automatically set to be at the common ideal circuit ground. Differential, on the other hand, means that the potential at either end of the port is defined by its position on the model. Hence, it allows us to chose the ground planes of the CPW on the chip as the reference ground of the differential port. Figure 4.15(c) shows part of the coplanar waveguide on the chip and the differential ports between ground plane and central conductor. The ports are arranged periodically with a distance of 100 μm along the waveguide in each gap reproducing the experimental setup. On the right of Fig. 4.15(c), the schematic of the electric circuit equivalent of half the rod is shown. This circuit is included as a SPICE model in between two pins of the differential port as indicated in Fig. 4.15(c). SPICE, Simulation Program with Integrated Circuit Emphasis, is "a general-purpose circuit simulation program for nonlinear dc, nonlinear transient, and linear ac analyses" [Qua+].

The SPICE model describes half of one rod, which leads from the central conductor across one gap to the ground plane and the term semi-rod will be used for it throughout the text. The parameters used in the model are given in Tab. 4.3. The value of the capacitance to central conductor C_c is slightly less than the value used for the calculation as given in Tab. 3.4. This improves the agreement between simulation and calculation, i.e. measurement. The calculation (cf. previous subsection) uses a simple two port lumped element transmission line model. Therefore, the slightly different values for C_c are not a big surprise, considering that the 3D environment influences the effective capacitance of the structure. As inductance of the dc-SQUID, the zero field inductance is used.

I_c [μA]	L_{j0} [pH]	R [Ω]	C_{in} [pF]	L_{geo} [pH]	$C_c/2$ [pF]
3.2	103	950	0.13	115	3.0

Table 4.3: Parameters of the semi-rod structures included in the differential port as shown in Fig. 4.15(c). The values are the same as given in Tab. 3.4, except for the capacitance C_c.

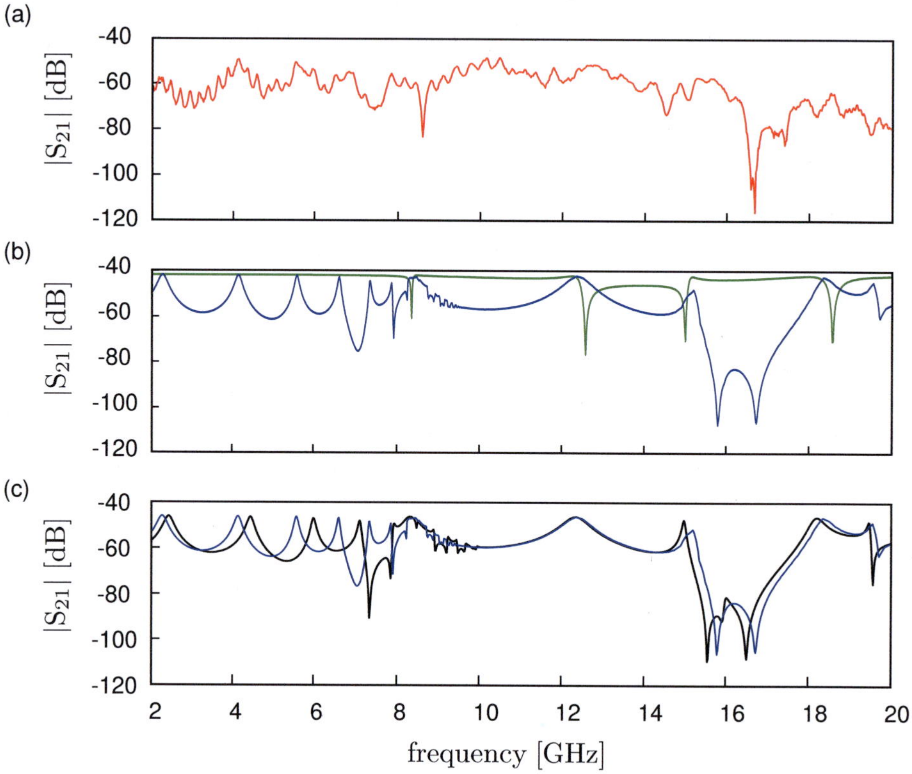

Fig. 4.16: Comparison of measured transmission magnitude with simulation results. (a) Measurement at $\Phi_{e0}/\Phi_0 = 0$. (b) Simulation result with 54 semi-rods included between the two pins of a differential port (blue). Simulation result with an empty waveguide (green). (c) The same simulation result using differential ports as above (blue). Simulation results using the simplified lumped element model introduced in the text (black).

The agreement between the measured result of the transmission magnitude at zero flux in Fig. 4.16(a) and the simulated result (blue) in Fig. 4.16(b) is clearly visible although not perfect. The simulation also shows the periodicity due to the finite size of the metamaterial. But, like the measurement, it does not display the stop band. Instead, the transmission magnitude stays at a high level. Additionally, the peaks at around 8 GHz, 12 GHz and 15 GHZ, as well as the double dip around 16 GHz are also found in the measured result, although not as pronounced and slightly shifted with respect to frequency. The shift can be explained by details in the real setup which the model does not reproduce.

For example, the position of the input and output port (red lines in Fig. 4.15(b)) influences the position of the peaks and the double resonance. The chip height in the experiment is also not exactly defined due to the dielectric glue between chip and PCB and an uncertainty in the milling of the hole. Nonetheless, considering the complexity of the model, it reproduces the experimental result reasonably well. However, the question why no stop band is observed still remains.

The green curve in Fig. 4.16(b), represents the simulated transmission magnitude through the empty CPW, i.e without rods. The comparison of green and blue line in Fig. 4.16(b) clearly shows that the transmission through the empty CPW exhibits dips, i.e. parasitic resonances, at the position where the transmission through the rod loaded waveguide shows peaks.

We will employ 3D field monitors at the frequency of dip and respective peak, e.g. 12.5 GHz, to show that both, dip and peak, are due to the same mode. These field monitors calculate the 3D field distribution of electric field, magnetic field and corresponding surface currents. Furthermore, we will the use the field monitors to demonstrate that the finite transmission in the frequency band of the stop band is due to parasitic transmission through the sample holder. Additionally, the field distribution indicates a possible reason for the different line shapes of peaks, which seem to be of Lorentzian shape, and dips, which look like Fano resonances.

The initial model containing the rods as SPICE elements between pins of differential ports cannot be used to get the 3D field distribution. Including the semi-rod as SPICE model requires a co-simulation of the circuit in addition to the finite element simulation. This configuration prohibits the use of the 3D field monitors. As replacement for the SPICE model, the rods have to be reduced and simplified to a RLC series element so that they can be included as lumped element ports. In order to obtain this RLC lumped element circuit, first of all, the intrinsic capacitances in the RCSJ model of single junction and dc-SQUID are neglected. Next, the leftover inductance L_p and resistance R_p of the single junction are turned from a parallel to a series circuit defined by L_s and R_s at a frequency of 10 GHz. The resulting values are given in Tab. 4.4. The corresponding values for the dc-SQUID are half the values of the single junction. Note that the transformation is correct only at this frequency but the deviations especially of the prominent inductive part are small.

Finally, the values of the full lumped element circuit are the total capacitance $C_{s,\mathrm{tot}}$ comprising the capacitance of one rod arm to ground and to the central conductor, the total inductance $L_{s,\mathrm{tot}}$ which includes the L_{geo} and L_s of single junction and dc-SQUID and the total resistance $R_{s,\mathrm{tot}}$ consisting of the series value R_s of single junction and dc-SQUID. The resulting values are given in Tab. 4.4. In Fig. 4.17, the frequency dependences of real (inset) and imaginary parts of the impedance of the rod, Z_{rod}, as given by Eq. (2.68) and impedance of the simplified RLC series circuit, Z_{LE}, are compared. The blue line shows the frequency dependence of Z_{rod}, while the black curve describes Z_{LE}.

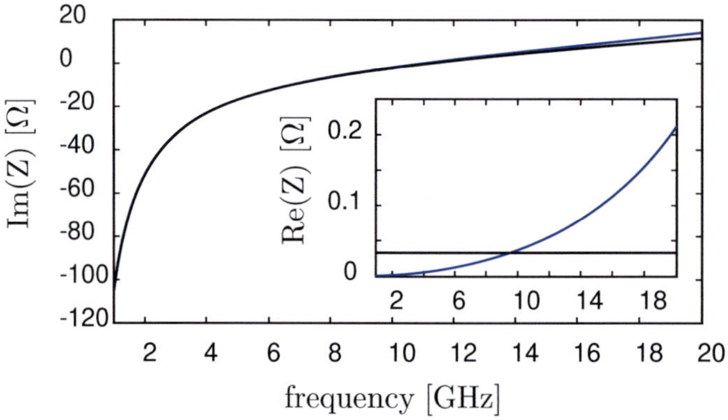

Fig. 4.17: Comparison of the imaginary and real (inset) part of the impedance Z_{rod} (blue) as given by Eq. (2.68) and the impedance of the simplified RLC series circuit (black).

Since R_s and L_s are calculated at a fixed frequency, the real part of Z_{LE} (inset) is constant unlike the real part of Z_{rod} (inset). This discrepancy, however, does not have a strong effect on the result, since the real part is small compared to the imaginary part. The deviation of the two imaginary parts at high frequencies is the result of neglecting the intrinsic Josephson capacitance. Overall, the agreement is sufficiently good to use the simplified structure instead of the rods in the simulation.

L_s [pH]	R_s [Ω]	$C_{s,\mathrm{tot}}$ [pF]	$L_{s,\mathrm{tot}}$ [pH]	$R_{s,\mathrm{tot}}$ [Ω]
103	0.04	0.75	269	0.06

Table 4.4: Parameters of the simplified RLC lumped element circuit. The description is found in the main text. L_s and R_s are the values for the series equivalent of the single Josephson junction neglecting the intrinsic capacitance.

The simulation result of $|S_{21}|$ for rods (blue) and RLC circuit (black) are shown in Fig. 4.16(c). Although the lumped element simulation does not reproduce the low frequency features accurately, the general shape of the transmission magnitude is reproduced. The agreement is best between 10 GHz and 14 GHz.

With the lumped element model, it is now possible to use the field monitors at specific frequencies and find the corresponding 3D field distribution for the case of the loaded and the unloaded waveguide. As frequencies for the field monitors, the resonance dips and peaks of the empty and loaded CPW are employed. Since the agreement between rods and series circuit is best at that frequency, the resonance dip of unloaded line (green) at 12.6 GHz and the respective peak of the loaded waveguide (black) at 12.3 GHz are chosen.

4.2.3 Simulated Field Distribution at 12.3 GHz and 12.6 GHz

Using the field monitors in the CST simulation, the 3D electric and magnetic field distribution is simulated as well as the surface currents induced by the fields. In order to understand the difference between the unloaded and the loaded waveguide, the amplitude of the absolute value of the surface currents offers the clearest picture.

The result of the simulation of the unloaded sample is presented in Fig. 4.18. The amplitude of the absolute value of the surface current between the input and output port (red lines) is illustrated in color scale. Figure 4.18(a) shows the top view of the PCB together with the waveguide on the chip. The inset indicates what part of the PCB is depicted for orientation. Since only dielectric material appears transparent in the figure, Fig. 4.18(b) shows the same view but with the chip removed. Now, the surface currents in the lower copper plane of the PCB are visible.

In the same way, the simulation result of the waveguide loaded with the RLC replacement of the rods is given in Fig. 4.19. Again, the amplitude of the surface current is shown once with (Fig. 4.19(a)) and once without (Fig. 4.19(b)) displayed chip. In both plots, the shape of the resonance in the sample holder, as seen in (b) of both figures, is very similar. It is the same mode, however with a slight asymmetric deformation in the case of the unloaded waveguide.

Considering the surface currents in the waveguide of the unloaded sample first, a current node in the middle of left waveguide section on the PCB indicates that a standing wave is created. The current anti-node on the chip feeds the resonance. Since the standing wave is not symmetric with respect to the

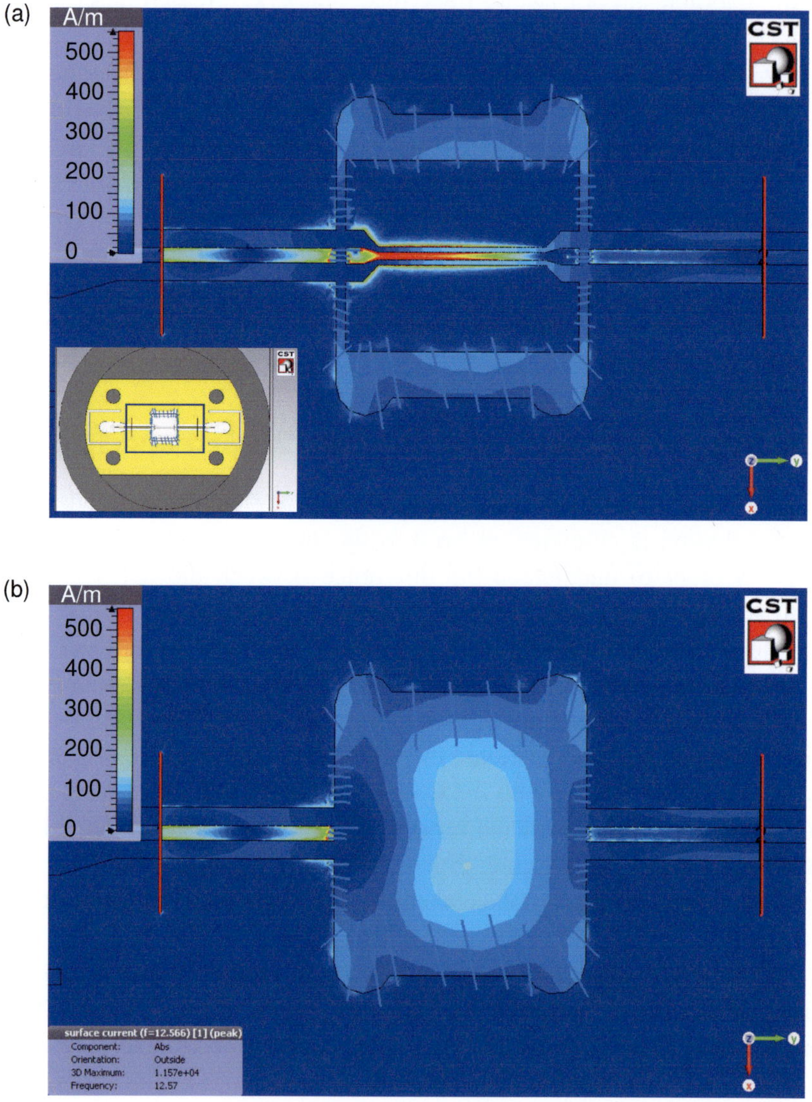

Fig. 4.18: Spatial distribution of the absolute value of the amplitude of the surface current density on the PCB and the unloaded waveguide on the chip at 12.6 GHz. (a) Top view on PCB and chip. The inset indicates which area of the sample holder is shown. (b) Same picture as in (a) but with removed chip to uncover the current distribution below.

(a)
(b)

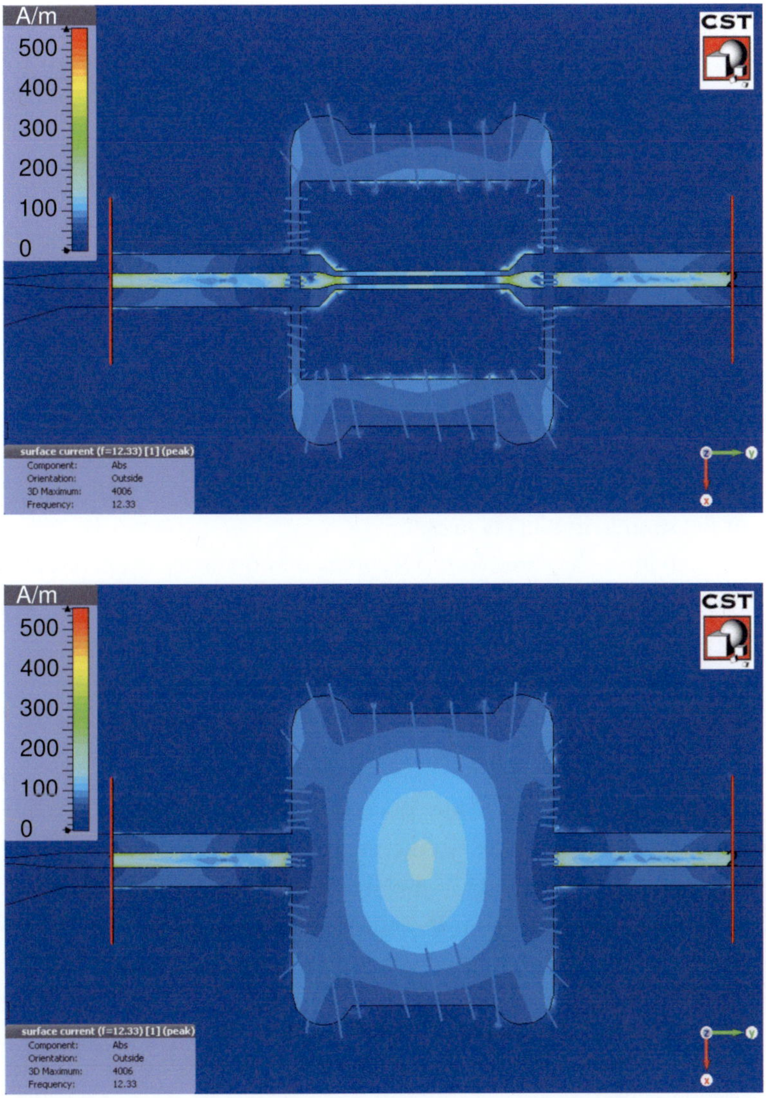

Fig. 4.19: Spatial distribution of the absolute value of the amplitude of the surface current density on the PCB and the waveguide on the chip loaded with rods (not shown) at 12.3 GHz. (a) Top view on PCB and chip. (b) Same picture as in (a) but with removed chip to uncover the current distribution below.

chip center, this asymmetry is reflected in the shape of the resonance mode. The coupling of the two resonances, the one in the waveguide and the one in the sample holder, yields a Fano like resonance [Fan61] and explains the slight frequency shift between the position of the dip (12.6 GHz) of the unloaded and the peak (12.3 GHz) of the loaded waveguide.

While the situation is similar in the periodically loaded sample, there is no standing wave pattern (no node) visible. Due to the stop band, which does not allow the wave to propagate far into the loaded transmission line, the waveguide on the chip is effectively split in half. This creates two waveguide stumps that act as antennas. The one on the left feeds the resonance, while, due to symmetry, the antenna on the right couples energy out of the resonance. Therefore, transmission from port 1 on the left hand side to port 2 on the right hand side is finite but not by means of the waveguide on the chip. Instead, the detour through a sample holder resonance is taken. The Lorentzian line shape of the peaks in the blue curve shown in Fig. 4.16(b) arises since only one resonance (that of the sample holder) is present.

This behavior is the same for all resonance frequencies observed in the green curve in Fig. 4.16(b). The resonant mode is always located in or around the hole in the PCB that contains the chip, its structure, however, differs.

simulation result of sample holder with hole

Due to the location of the resonance in the PCB, it seems to be a reasonable idea that by removing all metal below the chip, the resonances are deprived of the basis of their existence. In the simulation model this is easily done: A octagonal hole is "milled" through the PCB and the bottom part of the sample holder. The chip is now supported by the PCB only at its corners. Figure 4.20 displays a 3D view of the bottom part of the sample holder with the hole but without the chip.

The simulated transmission magnitude for the waveguide that uses the SPICE model of the rods, and not the lumped element description, is shown by the red line in Fig. 4.21. Compared to the sample holder without hole (blue), less resonances (peaks) are present. A double resonance at $\sim$ 12.5 GHz, however, is still present and prevents the appearance of the stop band.

By comparison of the 3D field structure of the resonance modes, it can be shown that the two maxima of the double peak correspond to the modes that are located at 8.3 GHz and 12.3 GHz of the sample holder without hole. The mode corresponding to higher frequency peak is about identical to the mode at

12.3 GHz, while lower frequency peak mode comprises features of the mode at 8.3 GHz and of the mode at 12.3 GHz.

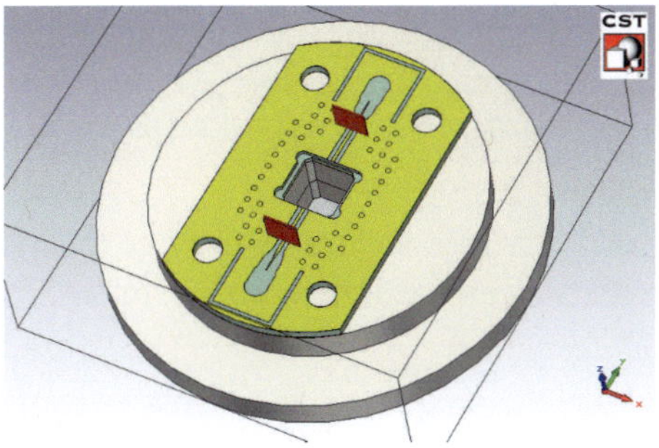

Fig. 4.20: Bottom part of the sample holder with octagonal hole below the chip position. The chip is not shown since it would cover the hole.

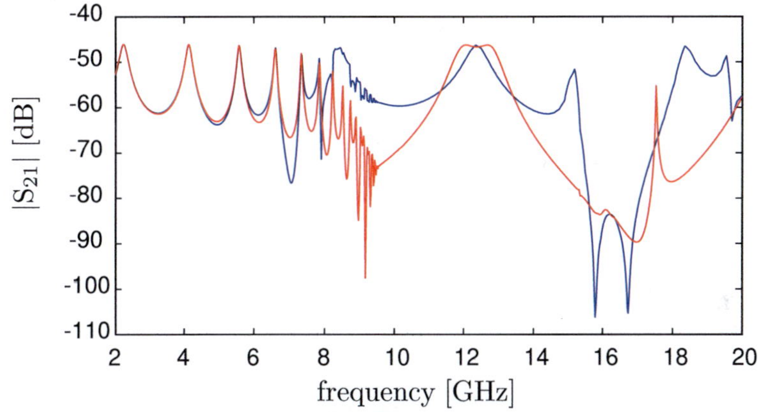

Fig. 4.21: Simulation results of the rod loaded transmission line. The result obtained from the sample holder with hole is shown by the red curve. For comparison the blue curve depicts the known result obtained without hole.

Shifting our focus to the pass band below the supposed onset of the stop band, we observe that the transmission decreases as the frequency increases and approaches the stop band. Exactly at the onset of the stop band, the transmission starts to rise again. This is consistent with the claim above. Once the frequency is reached, where waves cannot propagate anymore, the waveguide starts to behave like two antennas, opening the door for the detour.

4.2.4 Magnetic Flux Dependence

As a last point, the magnetic flux dependence of the rod loaded waveguide shall be discussed. This is of special interest, since the upper cut-off frequency of the pass band decreases in frequency as the flux is tuned away from zero. Hence, the onset of the stop band moves away from the lowest parasitic resonance and is less affected by it.

Figure 4.22(a) shows the frequency dependence of the measured transmission magnitude (uncalibrated) at flux values $\Phi_{e0}/\Phi_0 = -0.4$ (blue), -0.45 (red) and -0.47 (black). Unfortunately, the quality of the data suffers from the small period oscillations along the frequency axis which are most probably due to reflections in the setup. Nonetheless, two main features are visible. First, above 9 GHz, the plots do not change with changing flux anymore except around the resonance at 16.6 GHz. At this latter frequency, the second pass band crosses between $-0.48 < \Phi_{e0}/\Phi_0 < -0.45$ (cf. Fig. 4.13(c)). Second, the onset of the stop band moves to lower frequencies, leaving an increasing frequency band between this onset and the lowest parasitic resonance at 10 GHz.

The same behavior is observed more clearly when considering the simulation data. The inductance of the dc-SQUID is calculated for the above three flux values and is then implemented in the SPICE model. The result for $|S_{21}|$ is plotted in Fig. 4.22(b) in the corresponding colors. Figure 4.22(c) presents the same results but obtained using the CST model with the hole below the chip (cf. Fig. 4.20).

Both figures exhibit the decreasing onset of the stop band as the flux approaches $-\Phi_0/2$. The further this onset is separated from the lowest parasitic resonance frequency, the more the transmission at the onset decreases. The data also clearly shows how the transmission through the sample holder, which is magnetic field independent, takes over as soon as the upper cut-off frequency of the pass band is reached.

Additionally, the calculated transmission magnitude is shown at the same flux values in Fig. 4.22(d). The calculation uses the ideal two port transmission line model. The blue line ($\Phi_{e0}/\Phi_0 = -0.4$) does not reappear at higher frequencies, since the second pass band has not yet moved to below 20 GHz at that flux value. The comparison between calculation and simulation demonstrates clearly, how the sample holder resonance counteracts the stop band.

The stop band, however, is necessary in order to observe a NIR pass band once electric and magnetic meta-atoms are combined (cf. Sec. 2.4.4). As long

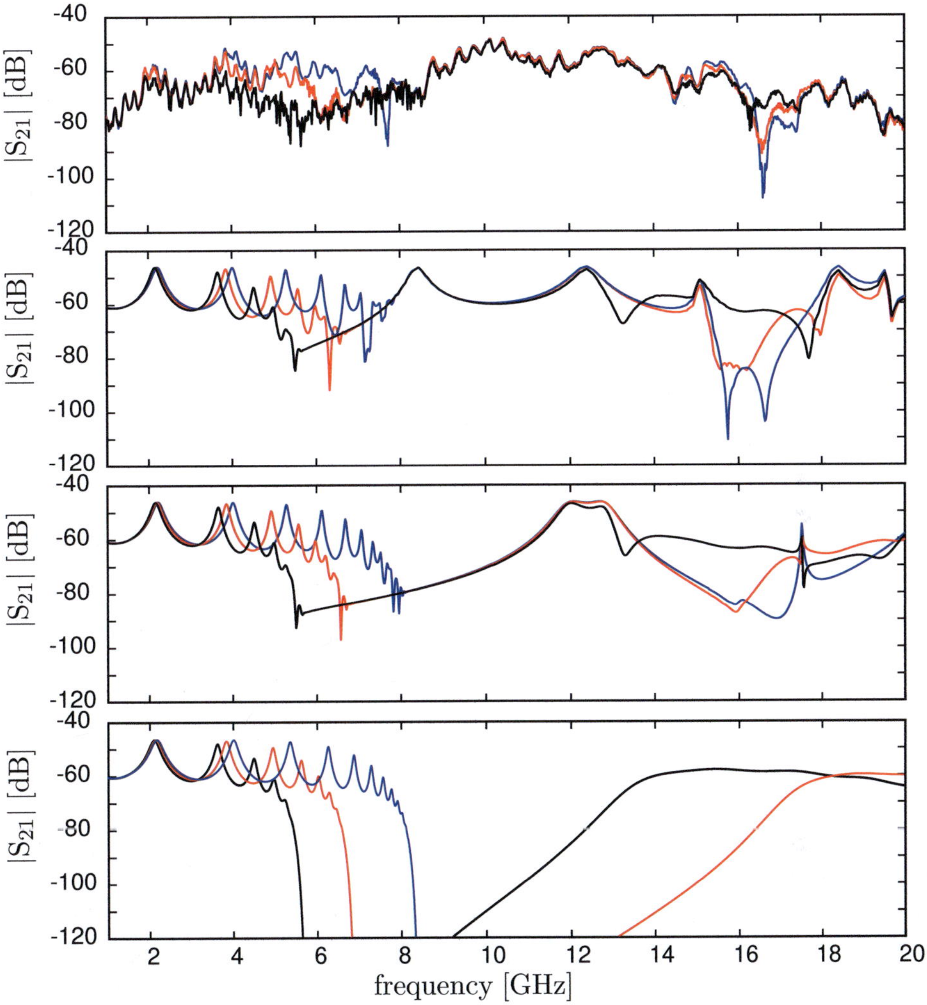

Fig. 4.22: (a) Measured transmission magnitude at flux values $\Phi_{e0}/\Phi_0 = -0.4$ (blue), -0.45 (red) and -0.47 (black). Simulated transmission magnitude at the same flux values and in corresponding colors is shown as obtained with (b) the model without hole below the chip (c) the model containing the hole. (d) Calculated transmission magnitude using the transmission line approach discussed above in Sec. 4.2.1

as the parasitic transmission through the sample holder is larger than the transmission in the NIR pass band of the composite medium, the parasitic transmission will be the dominant process. This changes however, if the situation is reversed. When the parasitic transmission is low enough, the tunable NIR pass band will be the dominant process and, hence, should be observable. For the model with the hole, the transmission shown in Fig. 4.22(c) goes down to $-90\,\mathrm{dB}$ for $\Phi_{e0}/\Phi_0 = -0.45$. This value may be sufficiently low so that the predicted NIR pass band can occur. However, it may appear only in the narrow frequency range between the onset of the stop band and the frequency, where the parasitic transmission has increased enough to become the dominant process again.

4.2.5 Discussion & Conclusion

The presented transmission data, measured on a waveguide that couples electrically to rod-like structures, partially confirm and partially disagree with the behavior predicted by a simple transmission line model. While the low frequency dependence of the transmission magnitude in the pass band was consistent with expectations, the anticipated stop band was not observable.

The same result was obtained from simulations, which pinpointed the nonexistent stop band to parasitic transmission through the sample holder. First investigations in the matter showed that alterations to the sample holder design should improve the situation and, if a suitable design was found, may finally overcome this problem.

Using the field monitors of the simulation at the parasitic resonances, the 3D field distribution with and without electric structures was determined. It showed that at the stop band, the microwave traveling along the waveguide was strongly attenuated on the length scale of one or two unit cells. The question arises if by also including magnetically coupled structures, this penetration is sufficient to create a negative index "material", which would then allow transmission through the composite medium. In this context, it may be sensible to develop an improved rod design, that couples less strongly to the waveguide, hence decreasing the impedance mismatch at the interface of the $50\,\Omega$ transmission line and loaded waveguide.

Nonetheless, with this investigation of a waveguide loaded with capacitively coupled structures, the first milestone for the development of a tunable superconducting negative index material was set.

5 Conclusion & Outlook

This Chapter concludes the thesis. It offers a brief summary of what has been achieved and how the results are interpreted. Additionally, it includes suggestions about further improvements that could be implemented and what outcome could be expected.

This thesis presents a novel approach to the topic of tunable metamaterials. The resonant behavior of the commonly used meta-atom, e.g. the split ring resonator, imposes a strong limitation. Specific values, for example zero or negative values, of the magnetic permeability are achievable only in a narrow frequency band. By including tunable elements into the resonator, its resonance frequency and thus the magnetic permeability of the material becomes tunable.

In this work, the Josephson junction was employed as tunable inductor in superconducting meta-atoms. The meta-atoms were included into a coplanar waveguide, creating a one-dimensional metamaterial. Two different kinds of meta-atoms were introduced. The rf-SQUID, a superconducting loop interrupted by a single Josephson junction with a shunt capacitance in parallel, was employed as magnetic meta-atom. It was placed in the gap of the coplanar waveguide, oriented with its area perpendicular to the magnetic field in the gap. The electric meta-atom couples capacitively to the waveguide and contains a dc-SQUID as tunable element in addition to a single junction. The resonance frequency of both meta-atoms was tunable by a constant magnetic field perpendicular to the (rf-/dc-) SQUID loop. By measuring the complex transmission coefficient with a vector network analyzer, the frequency and magnetic flux dependent transmission properties of the metamaterial were investigated.

The rf-SQUID, i.e. the magnetic meta-atom, has a resonance frequency that is tunable over a range of 5 GHz (30 % of the central frequency of 12 GHz) by applying a magnetic field in situ as shown in Sec. 4.1. On the other hand, its sensitivity to magnetic field also posed the biggest challenge for the implementation of a SQUID metamaterial with a collective resonance. In Sec. 3.3, measures were discussed that are necessary to protect the magnetic metamaterial from stray magnetic fields and the occurrence of Abrikosov vortices. Only an optimized experimental setup and sample design permitted the collective response of almost all meta-atoms in the coplanar waveguide. Once a collective resonance of the magnetic metamaterial was achieved, the corresponding transmission data was used to extract its frequency and flux dependent effective, relative magnetic permeability. Therefore, in Sec. 2.4.2, a novel technique, inspired by Nicholson, Ross and Weir, was developed that allows the retrieval of the relative magnetic permeability from only the transmission data. Using this technique together with a plausibility test suggested by Weir, the tunability of the magnetic permeability was clearly demonstrated. By decreasing the temperature, thus improving the quality factor, it was even possible to achieve negative values for the permeability in an appreciable frequency range.

Further work on this material has to aim at widening the frequency range where the permeability is less than zero. Therefore, the design of the sample needs to be adjusted to achieve a better quality factor. This should go hand in hand with an even more precise fabrication process, to further decrease the spread of parameters of the SQUIDs. Additionally, a quality factor that shows a monotonous dependence on flux between zero and half flux quantum has to be ensured. Only then a consistent range of tunability of the magnetic permeability is guaranteed within the minimal and maximal resonance frequency. Therefore, a sample holder needs to be developed without parasitic resonances in the frequency range of interest.

Parasitic resonances in the sample holder were also the main challenge that had and still has to be overcome in order to observe a stop band in the electric metamaterial discussed in Sec. 4.2.

The measured transmission data through a coplanar waveguide loaded with electrically coupling structures showed good agreement with the expected frequency and flux dependent behavior detailed in Sec. 2.4.3 but only in the low frequency pass band. The predicted stop band was not observed. Without an observable stop band or at least sufficiently low transmission in the electric metamaterial, the realization of a composite medium of tunable magnetic meta-atoms and electric structures was not possible.

Finite element simulations of the full 3D system indicated that the finite transmission in the stop band is due to parasitic transmission through the sample holder. The simulations also showed that by "pulling the rug out" from under the resonance, i.e. by removing any metallic material below the chip, the effect of the sample holder resonances is decreased. Resonances at lower frequencies were removed or rather pushed to higher frequencies. However, the resonances at higher frequencies still affected the stop band. On the other hand the parasitic effect was decreased the further the pass band was pushed to lower frequencies as the flux approaches odd integers of half flux quanta. It may be possible to observe the expected tunable NIR pass band of the composite medium at these low frequencies and flux values close to half flux quantum.

Samples containing a composite one-dimensional tunable metamaterial consisting of both kinds of meta-atoms already exist. Hence, improving the sample holder by milling a hole through PCB and bottom part at the position of the chip may in fact lead to an observable tunable NIR pass band for flux values close to $\Phi_0/2$ of the dc-SQUID in the rod structure. An experiment that may be carried out in the not so distant future.

Zusammenfassung & Ausblick

Diese Dissertation stellt einen neuen Ansatz zum Thema stimmbare Metamaterialien vor. Das resonante Verhalten gemeinhin gebräuchlicher Meta-Atome, z.B. des Split-Ring-Resonators, stellt eine starke Einschränkung dar. Bestimmte Werte, wie zum Beispiel Null oder negative Werte, nimmt die magnetische Permeabilität eines Materials bestehend aus Split-Ring-Resonatoren nur in einem engen Frequenzband an. Indem man nun stimmbare Elemente in den Resonator einfügt, wird seine Resonanzfrequenz und damit auch die magnetische Permeabilität des betreffenden Materials einstellbar.

In dieser Arbeit wird ein Josephson Kontakt als stimmbare Induktivität in supraleitenden Meta-Atomen verwendet. Die Meta-Atome sind in einen ebenen Wellenleiter eingebettet und stellen so ein eindimensionales Metamaterial dar.

Zwei unterschiedliche Arten von Meta-Atomen wurden vorgestellt. Das rf-SQUID ist ein supraleitender Ring unterbrochen von einem einzelnen Josephson Kontakt parallel geschaltet mit einem Kondensator und wird als magnetisches Meta-Atom verwendet. Es befindet sich im Schlitz eines koplanaren Wellenleiters und liegt mit seiner Fläche senkrecht zu den magnetischen Feldlinien im Schlitz. Die elektrischen Meta-Atome koppeln kapazitiv an den Wellenleiter und enthalten ein dc-SQUID als stimmbares Bauelement zusätzlich zu einem einzelnem Josephson Kontakt. Die Resonanzfrequenz beider Meta-Atome ist stimmbar mittels Anlegen eines konstanten magnetischen Feldes senkrecht zum Ring des (rf/dc)-SQUIDs. Die frequenz- und flussabhängigen Transmissionseigenschaften des jeweiligen Metamaterials wurden durch das Messen des komplexen Transmissionskoeffizienten mit Hilfe eines Vektornetzwerkanalysators untersucht.

Das rf-SQUID, d.h. das magnetische Meta-Atom, besitzt eine Resonanzfrequenz, die über einen Bereich von 5 GHz (entspricht 30 % der mittleren Frequenz von 12 GHz) durch Anlegen eines magnetischen Feldes stimmbar ist, wie in Abschn. 4.1 gezeigt wurde. Auf der anderen Seite stellt seine Empfindlichkeit gegenüber magnetischen Feldern gleichzeitig die größte Herausforderung bei der Verwirklichung eines SQUID-Metamaterials mit einer kollektiven Resonanz dar. In Abschn. 3.3 wurden die Maßnahmen diskutiert, die notwendig waren um das magnetische Metamaterial gegen magnetische Störfelder und das Auftreten von Abrikosov Vortizes zu schützen. Nur mit einem optimierten experimentellen Aufbau und Probendesign konnte die gemeinsame Anregung fast aller Meta-Atome im Wellenleiter verwirklicht werden.

Nachdem eine kollektive Resonanz des Metamaterials erzielt war, wurden die entsprechenden Transmissionsdaten dazu verwendet um die frequenz- und flussabhängige, relative magnetische Permeabilität zu extrahieren. Dazu wurde in Abschn. 2.4.2 eine neue Technik, inspiriert von Nicholson, Ross und Weir, vorgestellt, die es ermöglichte die relative magnetische Permeabilität nur aus den Transmissionsdaten zu gewinnen. Mit dieser Technik, in Verbindung mit einem von Weir vorgeschlagenen Plausibilitätstest, konnte die Stimmbarkeit der magnetischen Permeabilität eindeutig demonstriert werden. Durch Erniedrigen der Temperatur wurde außerdem der Qualitätsfaktor der Resonanz verbessert. Damit war es sogar möglich, in einem endlichen Frequenzbereich negative Werte für die Permeabilität zu erzielen.

Jegliche weitere Arbeit zu diesem Material sollte darauf abzielen, den Frequenzbereich, in dem die Permeabilität negativ ist, aufzuweiten. Dafür muss das Design der Probe angepasst werden um einen besseren Qualitätsfaktor zu erreichen. Parallel dazu sollte durch einen noch präziseren Fabrikationsprozess die Streuung der Parameter der SQUIDs, wie zum Beispiel der kritische Strom des Kontaktes, weiter verringert werden. Zusätzlich muss gewährleistet sein, dass der Qualitätsfaktor eine monotone Flussabhängigkeit zwischen Null und einem halben Flussquant aufweist. Nur dann kann die gleichmäßige Stimmbarkeit der magnetischen Permeabilität im gesamten Bereich zwischen minimaler und maximaler Resonanzfrequenz der SQUIDs sichergestellt werden. Um dies zu erreichen muss eine Probenhalter entwickelt werden, der keine parasitären Resonanzen im relevanten Frequenzbereich aufweist.

Parasitäre Resonanzen des Probenhalters stellten auch die größte zu bewältigende Herausforderung dar auf dem Weg hin zu einem beobachtbaren Stopband im elektrischen Metamaterial, das in Abschn. 4.2 behandelt wurde.

Die gemessene Transmission durch den ebenen Wellenleiter mit elektrisch koppelnde Strukturen ist in guter Übereinstimmung mit dem in Abschn. 2.4.3 behandelten frequenz- und flussabhängigen Verhalten, aber nur im Durchlassbereich bei niedrigeren Frequenzen. Das vorhergesagte Stopband wurde nicht beobachtet. Ohne ein beobachtbares Stopband oder zumindest ausreichend niedriger Transmission im elektrischen Metamaterial, ist die Umsetzung eines kombinierten Mediums aus stimmbaren magnetischen und elektrischen Meta-Atomen nicht möglich.

Finite Elemente Simulationen des gesamten dreidimensionalen Systems deuteten darauf hin, dass die nicht verschwindende Transmission im Stopband auf parasitäre Transmission durch den Probenhalter zurückzuführen ist. Die Simulationen zeigten auch, dass der negative Einfluss der Probenhalterresonanzen verringert werden kann, indem das Metall unter dem Chip entfernt wird. Damit wird der Resonanz sozusagen der Boden unter den Füßen entzogen. Resonanzen im Probenhalter bei niedrigeren Frequenzen wurden dadurch beseitigt bzw. zu höheren Frequenzen verschoben. Allerdings beeinflussen die Resonanzen bei höheren Frequenzen nach wie vor das Stopband. Ihr parasitärer Effekt wurde jedoch kleiner, je weiter das obere Ende des Durchlassbereichs zu niedrigeren Frequenzen verschoben wurde. Das wurde erreicht, in dem magnetische Flusswerte nahe bei ungeradzahligen Vielfachen von halben Flussquanten angelegt wurden. Unter Umständen ist es möglich das, im kombinierten Medium erwartete, stimmbare Durchlassfrequenzband mit negativem Brechungsindex bei diesen niedrigen Frequenzen und Flusswerten nahe von halben Flussquanten zu beobachten.

Proben mit dem kombinierten, eindimensionalen und stimmbaren Metamaterial, das beide Arten von Meta-Atomen enthält, sind bereits hergestellt. Somit könnte das Anpassen des Probenhalters zu einer erfolgreichen Umsetzung eines stimmbaren Metamaterials mit einem stimmbaren negativem Brechungsindex, also zu einem stimmbaren Durchlassfrequenzband führen. Diese Experiment könnte in nicht allzu ferner Zukunft durchgeführt werden.

Bibliography

[AB70] A. Abragam and B. Bleaney. *Electron paramagnetic resonance of transition ions*. The international series of monographs on physics. Oxford: Clarendon Press, 1970.

[Abr57] A. Abrikosov. "Magnetic properties of superconductors of the second group". *Sov. Phys. - JETP (Engl. Transl.)* 1957.

[AE03] M. Antoniades and G. Eleftheriades. "Compact linear lead/lag metamaterial phase shifters for broadband applications". *IEEE Antenn. Wireless Propag. Lett.* **2**(1) 2003, pp. 103–106.

[Al07] A. Alù, M. G. Silveirinha, A. Salandrino, and N. Engheta. "Epsilon-near-zero metamaterials and electromagnetic sources: Tailoring the radiation phase pattern". *Phys. Rev. B* **75**(15) 2007, p. 155410.

[Bar09] R. Barends. "Photon-detecting superconducting resonators". PhD thesis. Delft, The Netherlands: Technical University Delft, 2009.

[BCS57] J. Bardeen, L. N. Cooper, and J. R. Schrieffer. "Theory of Superconductivity". *Phys. Rev.* **108**(5) 1957, pp. 1175–1204.

[BJVK90] J. Baker-Jarvis, E. Vanzura, and W. Kissick. "Improved technique for determining complex permittivity with the transmission/reflection method". *IEEE Trans. Microw. Theory Tech.* **38**(8) 1990, pp. 1096–1103.

[Boa+11] A. Boardman, V. Grimalsky, Y. Kivshar, S. Koshevaya, M. Lapine, N. Litchinitser, V. Malnev, M. Noginov, Y. Rapoport, and V. Shalaev. "Active and tunable metamaterials". *Laser Photon. Rev.* **5**(2) 2011, pp. 287–307.

[Bri53] L. Brillouin. *Wave propagation in periodic structures : electric filters and crystal lattices*. 2. ed. with corr. and add. Dover books on engineering and engineering physics. New York: Dover Publ., 1953. ISBN: 0-486-60034-3.

[BTW13] M. Bartsch, P. Thoma, and B. Wagner. *https://www.cst.com/Products/CSTMWS*. 2013.

[But+13a] S. Butz, P. Jung, L. V. Filippenko, V. P. Koshelets, and A. V. Ustinov. "A one-dimensional tunable magnetic metamaterial". *Opt. Express* **21**(19) 2013, pp. 22540–22548.

[But+13b] S. Butz, P. Jung, L. V. Filippenko, V. P. Koshelets, and A. V. Ustinov. "Protecting SQUID metamaterials against stray magnetic fields". *Supercond. Sci. Technol.* **26**(9) 2013, p. 094003.

[Cap09] F. H. Capolino, ed. *Metamaterials handbook.* Vol. theor: Theory and phenomena of metamaterials. Boca Raton, Fla. [u.a.]: CRC, 2009.

[CB04] J. Clarke and A. I. Braginski, eds. *The SQUID handbook.* Vol. 1: Fundamentals and technology of SQUIDs and SQUID systems. Weinheim: Wiley-VCH, 2004. ISBN: 3-527-40229-2; 978-3-527-40229-8.

[CCS10] Chen Huanyang, Chan C. T., and Sheng Ping. "Transformation optics and metamaterials". *Nat Mater* **9**(5) 2010, pp. 387–396.

[DCL06] C. Du, H. Chen, and S. Li. "Quantum left-handed metamaterial from superconducting quantum-interference devices". *Phys. Rev. B* **74**(11) 2006, p. 113105.

[DCL08] C. Du, H. Chen, and S. Li. "Stable and bistable SQUID metamaterials". *J. Phys.: Condens. Matter* **20**(34) 2008, p. 345220.

[Dol+06] G. Dolling, C. Enkrich, M. Wegener, C. M. Soukoulis, and S. Linden. "Low-loss negative-index metamaterial at telecommunication wavelengths". *Opt. Lett.* **31**(12) 2006, pp. 1800–1802.

[Dol+07] G. Dolling, M. Wegener, C. M. Soukoulis, and S. Linden. "Negative-index metamaterial at 780 nm wavelength". *Opt. Lett.* **32**(1) 2007, pp. 53–55.

[Eck] J. Eckert. *http://www.supratrans.de/en/home/*.

[Edw+08] B. Edwards, A. Alù, M. E. Young, M. Silveirinha, and N. Engheta. "Experimental Verification of Epsilon-Near-Zero Metamaterial Coupling and Energy Squeezing Using a Microwave Waveguide". *Phys. Rev. Lett.* **100**(3) 2008, p. 033903.

[EIK02] G. Eleftheriades, A. Iyer, and P. Kremer. "Planar negative refractive index media using periodically L-C loaded transmission lines". *IEEE Trans. Microw. Theory Tech.* **50**(12) 2002, pp. 2702–2712.

[Ele05] G. V. H. Eleftheriades, ed. *Negative refraction metamaterials : fundamental principles and applications*. [Piscataway, NJ]: IEEE Press, 2005. ISBN: 0-471-60146-2; 978-0-471-60146-3.

[ELT08] M. Eleftheriou, N. Lazarides, and G. P. Tsironis. "Magnetoinductive breathers in metamaterials". *Phys. Rev. E* **77**(3) 2008, p. 036608.

[Eno+02] S. Enoch, G. Tayeb, P. Sabouroux, N. Guérin, and P. Vincent. "A Metamaterial for Directive Emission". *Phys. Rev. Lett.* **89**(21) 2002, p. 213902.

[Erg+10] T. Ergin, N. Stenger, P. Brenner, J. B. Pendry, and M. Wegener. "Three-Dimensional Invisibility Cloak at Optical Wavelengths". *Science* **328**(5976) 2010, pp. 337–339.

[Fan+05] N. Fang, H. Lee, C. Sun, and X. Zhang. "Sub-Diffraction-Limited Optical Imaging with a Silver Superlens". *Science* **308**(5721) 2005, pp. 534–537.

[Fan61] U. Fano. "Effects of Configuration Interaction on Intensities and Phase Shifts". *Phys. Rev.* **124**(6) 1961, pp. 1866–1878.

[Fen12] S. Feng. "Loss-Induced Omnidirectional Bending to the Normal in ε-Near-Zero Metamaterials". *Phys. Rev. Lett.* **108**(19) 2012, p. 193904.

[FN13] V. Y. Fedorov and T. Nakajima. "All-angle collimation of incident light in ε-near-zero metamaterials". *Opt. Express* **21**(23) 2013, pp. 27789–27795.

[GBMM02] P. Gay-Balmaz, C. Maccio, and O. J. F. Martin. "Microwire arrays with plasmonic response at microwave frequencies". *Appl. Phys. Lett.* **81**(15) 2002, pp. 2896–2898.

[GE04] A. Grbic and G. V. Eleftheriades. "Overcoming the Diffraction Limit with a Planar Left-Handed Transmission-Line Lens". *Phys. Rev. Lett.* **92**(11) 2004, p. 117403.

[Gil+04] I. Gil, J. Garcia-Garcia, J. Bonache, F. Martin, M. Sorolla, and R. Marques. “Varactor-loaded split ring resonators for tunable notch filters at microwave frequencies”. *Electronics Letters* **40**(21) 2004, pp. 1347–1348.

[GJM08] A. Grbic, L. Jiang, and R. Merlin. “Near-Field Plates: Subdiffraction Focusing with Patterned Surfaces”. *Science* **320**(5875) 2008, pp. 511–513.

[Gub+05] A. I. Gubin, K. S. Il’in, S. A. Vitusevich, M. Siegel, and N. Klein. “Dependence of magnetic penetration depth on the thickness of superconducting Nb thin films”. *Phys. Rev. B* **72**(6) 2005, p. 064503.

[HBC03] A. A. Houck, J. B. Brock, and I. L. Chuang. “Experimental Observations of a Left-Handed Material That Obeys Snells Law”. *Phys. Rev. Lett.* **90**(13) 2003, p. 137401.

[Jac06] J. D. Jackson. *Klassische Elektrodynamik : [7 Tabellen]*. 4., überarb. Aufl. Berlin [u.a.]: de Gruyter, 2006.

[Jos65] B. Josephson. “Supercurrents through barriers”. *Adv. Phys.* **14**(56) 1965, pp. 419–451.

[Jun+13] P. Jung, S. Butz, S. V. Shitov, and A. V. Ustinov. “Low-loss tunable metamaterials using superconducting circuits with Josephson junctions”. *Appl. Phys. Lett.* **102**(6) 2013, p. 062601.

[Jun+14] P. Jung, S. Butz, M. Marthaler, M. V. Fistul, J. Leppäkangas, V. P. Koshelets, and A. V. Ustinov. “Multistability and switching in a superconducting metamaterial”. *Nat. Commun.* **5** 2014.

[Kan+08] L. Kang, Q. Zhao, H. Zhao, and J. Zhou. “Magnetically tunable negative permeability metamaterial composed by split ring resonators and ferrite rods”. *Opt. Express* **16**(12) 2008, pp. 8825–8834.

[Kap+11] P. V. Kapitanova, S. I. Maslovski, I. V. Shadrivov, P. M. Voroshilov, D. S. Filonov, P. A. Belov, and Y. S. Kivshar. “Controlling split-ring resonators with light”. *Appl. Phys. Lett.* **99**(25), 251914 2011.

[Kim+94] D. H. Kim, K. E. Gray, J. D. Hettinger, J. H. Kang, and S. S. Choi. "Resistive measurement of the temperature dependence of the penetration depth of Nb in Nb/AlOx/Nb Josephson junctions". *J. Appl. Phys.* **75**(12) 1994, pp. 8163–8167.

[Kit05] C. Kittel. *Introduction to solid state physics*. 8. ed. [Internat. ed.] New York: Wiley, 2005.

[KO11] H. Kamerlingh Onnes. "Further experiments with liquid helium. C. On the change of electric resistance of pure metals at very low temperatures etc. IV. The resistance of pure mercury at helium temperatures". *Proceedings* **13** 1911, pp. 1274–1276.

[Kro44] G. Kron. "Equivalent Circuit of the Field Equations of Maxwell-I". *Proceedings of the IRE* **32**(5) 1944, pp. 289–299.

[Lee+05] H. Lee, Y. Xiong, N. Fang, W. Srituravanich, S. Durant, M. Ambati, C. Sun, and X. Zhang. "Realization of optical superlens imaging below the diffraction limit". *New J. Phys.* **7**(1) 2005, p. 255.

[Lic91] K. K. Licharev. *Dynamics of Josephson junctions and circuits*. Philadelphia [u.a.]: Gordon and Breach, 1991.

[Lin+04] S. Linden, C. Enkrich, M. Wegener, J. Zhou, T. Koschny, and C. M. Soukoulis. "Magnetic Response of Metamaterials at 100 Terahertz". *Science* **306**(5700) 2004, pp. 1351–1353.

[LLI05] A. Lai, K. M. K. H. Leong, and T. Itoh. "A novel N-port series divider using infinite wavelength phenomena". In: *Microwave Symposium Digest, 2005 IEEE MTT-S International*. 2005.

[LLI07] A. Lai, K. M. K. H. Leong, and T. Itoh. "Infinite Wavelength Resonant Antennas With Monopolar Radiation Pattern Based on Periodic Structures". *IEEE Antenn. Wireless Propag. Lett.* **55**(3) 2007, pp. 868–876.

[LT07] N. Lazarides and G. P. Tsironis. "rf superconducting quantum interference device metamaterials". *Appl. Phys. Lett.* **90**(16), 163501 2007.

[LT13] N. Lazarides and G. P. Tsironis. "Multistability and self-organization in disordered SQUID metamaterials". *Supercond. Sci. Technol.* **26**(8) 2013, p. 084006.

[MO33] W. Meissner and R. Ochsenfeld. "Ein neuer Effekt bei Eintritt der Supraleitfähigkeit". *Naturwissenschaften* **21**(44) 1933, pp. 787–788.

[NR70] A. Nicolson and G. F. Ross. "Measurement of the Intrinsic Properties of Materials by Time-Domain Techniques". *IEEE Trans. Instrum. Meas.* **19**(4) 1970, pp. 377–382.

[Par+03] C. G. Parazzoli, R. B. Greegor, K. Li, B. E. C. Koltenbah, and M. Tanielian. "Experimental Verification and Simulation of Negative Index of Refraction Using Snells Law". *Phys. Rev. Lett.* **90**(10) 2003, p. 107401.

[Pen+98] J. Pendry, A. J. Holden, D. J. Robbins, and W. J. Stewart. "Low frequency plasmons in thin-wire structures". *Journal of Physics: Condensed Matter* **10**(22) 1998, p. 4785.

[Pen+99] J. Pendry, A. J. Holden, D. J. Robbins, and W. J. Stewart. "Magnetism from conductors and enhanced nonlinear phenomena". *IEEE Trans. Microw. Theory Tech.* **47**(11) 1999, pp. 2075–2084.

[Pen00] J. B. Pendry. "Negative Refraction Makes a Perfect Lens". *Phys. Rev. Lett.* **85**(18) 2000, pp. 3966–3969.

[Pie+07] S. V. Pietambaram, N. D. Rizzo, R. W. Dave, J. Goggin, K. Smith, J. M. Slaughter, and S. Tehrani. "Low-power switching in magnetoresistive random access memory bits using enhanced permeability dielectric films". *Appl. Phys. Lett.* **90**(14), 143510 2007.

[PNS07] Y. Polyakov, S. Narayana, and V. Semenov. "Flux Trapping in Superconducting Circuits". *IEEE Trans. Appl. Supercond.* **17**(2) 2007, pp. 520–525.

[Poz05] D. M. Pozar. *Microwave engineering*. 3. ed. Hoboken, NJ: Wiley, 2005. ISBN: 0-471-44878-8; 978-0-471-44878-5.

[Qua+] T. Quarles, D. Pederson, R. Newton, A. Sangiovanni-Vincentelli, and C. Wayne. *http://bwrcs.eecs.berkeley.edu/Classes/IcBook/SPICE/*.

[RA06] M. Ricci and S. Anlage. "Single superconducting split-ring resonator electrodynamics". *Appl. Phys. Lett.* **88**(26) 2006, p. 264102.

[RBLL95] P. S. Russell, T. A. Birks, and F. D. Lloyd-Lucas. "Photonic Bloch Waves and Photonic Band Gaps". In: *Confined Electrons and Photons*. Ed. by E. Burstein and C. Weisbuch. Vol. 340. NATO ASI Series. Springer US, 1995, pp. 585–633. ISBN: 978-1-4613-5807-7.

[Ric+07] M. Ricci, H. Xu, R. Prozorov, A. Zhuravel, A. Ustinov, and S. Anlage. "Tunability of Superconducting Metamaterials". *IEEE Trans. Appl. Supercond.* **17**(2) 2007, pp. 918–921.

[ROA05] M. Ricci, N. Orloff, and S. M. Anlage. "Superconducting metamaterials". *Appl. Phys. Lett.* **87**(3), 034102 2005.

[Sch97] V. V. Schmidt. *The physics of superconductors: Introduction to fundamentals and applications*. Berlin: Springer, 1997.

[SE06] M. Silveirinha and N. Engheta. "Tunneling of Electromagnetic Energy through Subwavelength Channels and Bends using ε-Near-Zero Materials". *Phys. Rev. Lett.* **97**(15) 2006, p. 157403.

[SFM04] G. Stan, S. B. Field, and J. M. Martinis. "Critical Field for Complete Vortex Expulsion from Narrow Superconducting Strips". *Phys. Rev. Lett.* **92**(9) 2004, p. 097003.

[Sha07] V. M. Shalaev. "Optical negative-index metamaterials". *Nat Photon* **1**(1) 2007, pp. 41–48.

[SK00] D. R. Smith and N. Kroll. "Negative Refractive Index in Left-Handed Materials". *Phys. Rev. Lett.* **85**(14) 2000, pp. 2933–2936.

[Smi+00] D. R. Smith, W. J. Padilla, D. C. Vier, S. C. Nemat-Nasser, and S. Schultz. "Composite Medium with Simultaneously Negative Permeability and Permittivity". *Phys. Rev. Lett.* **84**(18) 2000, p. 4184.

[Smi+05] D. R. Smith, D. C. Vier, T. Koschny, and C. M. Soukoulis. "Electromagnetic parameter retrieval from inhomogeneous metamaterials". *Phys. Rev. E* **71**(3) 2005, p. 036617.

[SMK06] I. V. Shadrivov, S. K. Morrison, and Y. S. Kivshar. "Tunable splitring resonators for nonlinear negative-index metamaterials". *Opt. Express* **14**(20) 2006, pp. 9344–9349.

[SPW04] D. R. Smith, J. B. Pendry, and M. C. K. Wiltshire. "Metamaterials and Negative Refractive Index". *Science* **305**(5685) 2004, pp. 788–792.

[SSS01] R. A. Shelby, D. R. Smith, and S. Schultz. "Experimental Verification of a Negative Index of Refraction". *Science* **292**(5514) 2001, pp. 77–79.

[Suc+13] H. Suchowski, K. OBrien, Z. J. Wong, A. Salandrino, X. Yin, and X. Zhang. "Phase MismatchFree Nonlinear Propagation in Optical Zero-Index Materials". *Science* **342**(6163) 2013, pp. 1223–1226.

[SZ67] A. H. Silver and J. E. Zimmerman. "Quantum States and Transitions in Weakly Connected Superconducting Rings". *Phys. Rev.* **157**(2) 1967, pp. 317–341.

[Tin04] M. Tinkham. *Introduction to superconductivity*. 2. ed. Mineola, NY: Dover Publ., 2004. ISBN: 0-486-43503-2; 978-0-486-43503-9.

[Tre+13] M. Trepanier, D. Zhang, O. Mukhanov, and S. M. Anlage. "Realization and Modeling of Metamaterials Made of rf Superconducting Quantum-Interference Devices". *Phys. Rev. X* **3**(4) 2013, p. 041029.

[Ves68] V. G. Veselago. "The electrodynamics of substances with simultaneously negative values of ε and μ". *Sov. Phys. USPEKHI* **10**(4) 1968, pp. 509–514.

[Wei74] W. B. Weir. "Automatic measurement of complex dielectric constant and permeability at microwave frequencies". *Proceedings of the IEEE* **62**(1) 1974, pp. 33–36.

[Wu+11] J. Wu, B. Jin, Y. Xue, C. Zhang, H. Dai, L. Zhang, C. Cao, L. Kang, W. Xu, J. Chen, and P. Wu. "Tuning of superconducting niobium nitride terahertz metamaterials". *Opt. Express* **19**(13) 2011, pp. 12021–12026.

[YA13] J.-H. Yeh and S. M. Anlage. "In situ broadband cryogenic calibration for two-port superconducting microwave resonators". *Rev. Sci. Instrum.* **84**(3), 034706 2013.

[Zee97] P. Zeeman. "On the influence of Magnetism on the Nature of the Light emitted by a Substance". *Philosophical Magazine* **43** 1897, p. 226.

[Zha+07] Q. Zhao, L. Kang, B. Du, B. Li, J. Zhou, H. Tang, X. Liang, and B. Zhang. "Electrically tunable negative permeability metamaterials based on nematic liquid crystals". *Appl. Phys. Lett.* **90**(1), 011112 2007.

[Zho+05] J. Zhou, T. Koschny, M. Kafesaki, E. N. Economou, J. B. Pendry, and C. M. Soukoulis. "Saturation of the Magnetic Response of Split-Ring Resonators at Optical Frequencies". *Phys. Rev. Lett.* **95**(22) 2005, p. 223902.

[Zio04] R. W. Ziolkowski. "Propagation in and scattering from a matched metamaterial having a zero index of refraction". *Phys. Rev. E* **70**(4) 2004, p. 046608.

List of Publications

1. P. Jung, S. Butz, S. V. Shitov and A. V. Ustinov. "Low-loss Tunable Metamaterials using Superconducting Circuits with Josephson Junctions". *Appl. Phys. Lett.* **102**, 062601 (2013).

2. S. Butz, P. Jung, L. V. Filippenko, V. P. Koshelets and A. V. Ustinov. "Protecting SQUID Metamaterials against Stray Magnetic Fields". *Supercond. Sci. Technol.* **26**, 094003 (2013).

3. S. Butz, P. Jung, L. V. Filippenko, V. P. Koshelets and A. V. Ustinov. "A One-Dimensional Tunable Magnetic Metamaterial". *Opt. Express* **21**, 22540 (2013).

4. S. Butz, P. Jung, L. V. Filippenko, V. P. Koshelets and A. V. Ustinov. "A One-Dimensional Tunable Magnetic Metamaterial: Erratum" *Opt. Express* **22**, 13041 (2014).

5. E. A. Ovchinnikova, S. Butz, P. Jung, V. P. Koshelets, L. V. Filippenko, A. S. Averkin, S. V. Shitov and A. V. Ustinov. "Design and Experimental Study of Superconducting Left-Handed Transmission Lines with Tunable Dispersion". *Supercond. Sci. Technol.* **26**, 114003 (2013).

6. P. Jung, S. Butz, M. Marthaler, M. V. Fistul, J. Leppkangas, V. P. Koshelets and A. V. Ustinov. "Multistability and Switching in a Superconducting Metamaterial" *Nat. Commun.* **5**, 3730 (2013).

7. S. Butz, A. K. Feofanov, K. G. Fedorov, H. Rotzinger, A. U. Thomann, B. Mackrodt, R. Dolata, V. B. Geshkenbein, G. Blatter and A. V. Ustinov. "Flux Dependent Crossover Between Quantum and Classical Behavior in a dc SQUID". *Phys. Rev. Lett.* **113**, 247005 (2014).

Acknowledgements

First of all, I want to thank Professor Alexey V. Ustinov. He suggested this interesting topic and put Philipp Jung at my side. During the, all in all, five years that I spent in his group, his advice and ideas have always been a valuable inspiration. The freedom with which he entrusted me in my work is also greatly appreciated.

Next, I want to acknowledge Professor Carsten Rockstuhl who kindly agreed to be the second referee of this thesis.

I thank Philipp Jung for the great collaboration, for bravely proof reading the whole thesis and most of all for his friendship. I can only hope that he appreciated our working together as much as I did. I am also very grateful to all the rest of my co-workers. Without you, the time in and outside lab and office would not have been so much fun. I especially want to thank Sasha Lukashenko for always having a solution to any technical problem, Hannes Rotzinger for always having the time for questions and a good answer, Martin Weides for always knowing a good citation and Jürgen Lisenfeld for proof reading part of this thesis. I also am happy to have had such great fellow PhD and Diploma/Master students. And I want to thank Sebastian Probst and Jochen Braumüller and many more in 4-10 and 4-5 for not being just collegues but also good friends.

I also want to acknowledge the collaboration with Valery P. Koshelets and his team at the Kotel'nikov Institute of Radio Engineering and Electronics (IREE RAS) in Moscow, Russia. They fabricated all the samples used in this work and helped adjust and improve our designs. In this context, I also appreciate the accomplishments of our in-house, mechanical and electronic workshops. Without their work, quick and necessary changes to the measurement setup could never have been realized within the given time constraints. Thanks also goes to the secretaries office and Lars Behrens. They take a lot of administrative work off our shoulders.

I thank the Landesgraduiertenfrderung Baden-Württemberg, who supported me financially with a scholarship for more then two and a half years.

My time in Karlsruhe has been a great time because of good friends I made, especially Christina and Tob. Thanks to all of you for the support and patience during the last few months but mostly for your existence and for your willingness to keep me company.

Last, but certainly not least, I want to thank my family. My parents never once doubted that I could do this and supported me in anything.

Experimental Condensed Matter Physics
(ISSN 2191-9925)

Herausgeber
Physikalisches Institut

Prof. Dr. Hilbert von Löhneysen
Prof. Dr. Alexey Ustinov
Prof. Dr. Georg Weiß
Prof. Dr. Wulf Wulfhekel

Die Bände sind unter www.ksp.kit.edu als PDF frei verfügbar
oder als Druckausgabe bestellbar.

Band 1 Alexey Feofanov
Experiments on flux qubits with pi-shifters. 2011
ISBN 978-3-86644-644-1

Band 2 Stefan Schmaus
Spintronics with individual metal-organic molecules. 2011
ISBN 978-3-86644-649-6

Band 3 Marc Müller
Elektrischer Leitwert von magnetostriktiven Dy-Nanokontakten. 2011
ISBN 978-3-86644-726-4

Band 4 Torben Peichl
Einfluss mechanischer Deformation auf atomare Tunnelsysteme – untersucht mit Josephson Phasen-Qubits. 2012
ISBN 978-3-86644-837-7

Band 5 Dominik Stöffler
Herstellung dünner metallischer Brücken durch Elektromigration und Charakterisierung mit Rastersondentechniken. 2012
ISBN 978-3-86644-843-8

Band 6 Tihomir Tomanic
Untersuchung des elektronischen Oberflächenzustands von Ag-Inseln auf supraleitendem Niob (110). 2012
ISBN 978-3-86644-898-8

Band 7 Lukas Gerhard
Magnetoelectric coupling at metal surfaces. 2013
ISBN 978-3-7315-0063-6

Experimental Condensed Matter Physics
(ISSN 2191-9925)

Band 8 Kirill Fedorov
Fluxon readout for superconducting flux qubits. 2013
ISBN 978-3-7315-0067-4

Band 9 Jochen Zimmer
Cooper pair transport in arrays of Josephson junctions. 2014
ISBN 978-3-7315-0130-5

Band 10 Oliver Berg
Elektrischer Transport durch Nanokontakte von Selten-Erd-Metallen. 2014
ISBN 978-3-7315-0209-8

Band 11 Grigorij Jur'evic Grabovskij
Investigation of coherent microscopic defects inside the tunneling barrier of a Josephson junction. 2014
ISBN 978-3-7315-0210-4

Band 12 Cornelius Thiele
STM Characterization of Phenylene-Ethynylene Oligomers on Au(111) and their Integration into Carbon Nanotube Nanogaps. 2014
ISBN 978-3-7315-0235-7

Band 13 Michael Peter Schackert
Scanning Tunneling Spectroscopy on Electron-Boson Interactions in Superconductors. 2014
ISBN 978-3-7315-0238-8

Band 14 Susanne Butz
One-Dimensional Tunable Josephson Metamaterials. 2014
ISBN 978-3-7315-0271-5